Praise for *Fut*

"Brad Allenby has written a terse, intense, and uncompromising summary of the threats posed by a range of disruptive technological innovations that affect our ability to wage contemporary warfare. Allenby's probing analysis and sensible advice concerning public policy alternatives (such as the reinstatement of universal conscription), when coupled with emerging ethical norms of responsible behavior in fields like robotics and cyberconflict, may effectively counteract the growing tendencies toward social dissolution and moral collapse in the face of these formidable technological challenges."

Professor Emeritus George R. Lucas, Jr.
U.S. Naval Academy
visiting professor, Notre Dame University
author of Ethics and Cyber Warfare *(2016)*

"Brad Allenby's new book, *Future Conflict & Emerging Technologies,* is required reading for anyone trying to implement strategy—financial, military, or otherwise—in our complex and rapidly changing world, characterized by competing companies and nations striving to further their interests even as they grapple with an unknowable, deeply challenging future and emerging technologies of unprecedented power."

James Hennessy
former President and CEO of ING Mutual Funds
current Chairman of Waterton Global Resource Management

"New technologies are adopted first by those for whom incremental advantage is critical, which is why in today's world military and security organizations are on the leading edge. As the pace of technological change increases exponentially, ISIS has figured out how to weaponize social media, China how to game entire societies, and Russia how to hack American elections - and everyone is trying to understand the military implications of gene editing tech. At such a time of dizzying upheaval, Brad Allenby in *Future Conflict & Emerging Technologies* brings imagination, wisdom, and much-needed perspective to the daunting but crucial task of ethically, rationally, and responsibly managing conflict and emerging technologies as we move into an uncertain and difficult future."

Professor Joel Garreau
Sandra Day O'Connor College of Law and affiliated faculty at the School for the Future of Innovation in Society
Arizona State University
former reporter and editor at The Washington Post
and author of Radical Evolution *(2005)*

The Rightful Place of Science: Future Conflict & Emerging Technologies

The Rightful Place of Science:

Future Conflict & Emerging Technologies

Braden R. Allenby

Consortium for Science, Policy & Outcomes
Tempe, AZ and Washington, DC

THE RIGHTFUL PLACE OF SCIENCE:
Future Conflict & Emerging Technologies

Printed in Charleston, South Carolina.

The Rightful Place of Science series explores the complex interactions among science, technology, politics, and the human condition.

For information on The Rightful Place of Science series,
write to: Consortium for Science, Policy & Outcomes
PO Box 875603, Tempe, AZ 85287-5603
Or visit: http://www.cspo.org

Other volumes in this series:

Cavalier, D. & Kennedy, E. B., eds. 2016. *The Rightful Place of Science: Citizen Science*. Tempe, AZ: Consortium for Science, Policy & Outcomes.

Benessia, A., Funtowicz, S., Giampietro, M., Guimarães Pereira, Â., Ravetz, J., Saltelli, A., Strand, R., and van der Sluijs, J. P. 2016. *The Rightful Place of Science: Science on the Verge*. Tempe, AZ: Consortium for Science, Policy & Outcomes.

Model citation for this volume:

Allenby, B. R. 2016. *The Rightful Place of Science: Future Conflict & Emerging Technologies*. Tempe, AZ: Consortium for Science, Policy & Outcomes.

ISBN: 0692774394

ISBN-13: 978-0692774397

LCCN: 2016914426

FIRST EDITION, SEPTEMBER 2016

CONTENTS

INTRODUCTION

The raison d'etre for the series of essays in this book is, simply put, the challenge of living in a world where the opportunities for conflict are rapidly multiplying, and where the accelerating evolution and democratization of military and security technologies make such conflicts far riskier. This rapidly evolving environment combines old verities and new, destabilizing geopolitical and technological systems in ways that make many familiar institutional patterns and behaviors questionable, if not obsolete.

This matters for one basic reason: conflict is inevitable for the foreseeable future. Violent conflict, however, in many cases may not be. If we can better understand current challenges and trends, perhaps we can reduce violence and social and cultural damage. This does not mean that we should expect to be able to eschew military or strategic engagement, but it does mean that we need a broader and more sophisticated framework appropriate for an increasingly complex, fast-moving, information-dense world. At the highest level, this requires developing a more robust theory of conflict in general, and emerging military and security technologies in particular.

The first two essays in this volume establish the context driving the need for a more comprehensive strategic framework. Chapter 1 discusses the changing geopolitical and strategic context within which conflict is occurring, using the examples of the Russian invasion of Crimea, the Chinese doctrine of "unrestricted warfare," and the rise of the so-called Islamic State, or ISIS. Each of these examples points to shifts in doctrine that adversaries of the United States are making in response to its dominance in conventional military capability. More fundamentally, the obsolescence of perceiving Western values as universal and the shift away from a Westphalian, state-based governance system both contribute significantly to the increasing complexity of the geopolitical environment.

Chapter 2 discusses emerging military and security technologies and some of the applied ethical issues that they raise. It also analyzes the drivers behind the development and deployment of these technologies, which strongly suggest that they will be difficult to stop or significantly restrict. Chapter 2 also provides a brief outline of some of the major elements of the laws, regulations, and norms that currently govern conflict, and which may already be contingent given technological and geopolitical trends.

The final three chapters, based on essays originally appearing in *Issues in Science and Technology Policy* and *The Bulletin of Atomic Scientists,* provide case studies illustrating some of these observations. Chapter 3, based on an article co-authored with Mark Hagerott, presents a situation where some of the implications of emerging technologies can be managed not by direct regulation of the technology itself, but by an apparently unrelated policy initiative: the reinstitution of universal conscription. Chapters 4 and 5 deal with two classes of technology, cyber and robotics, which taken together have a huge impact on the viability of current legal and regulatory structures governing conflict, as well as significant potential for disrupting current power relationships.

The reader should remember that these essays, even taken together, are not nearly comprehensive. In particular, while technology-specific case studies are useful for discussions of the viability of current norms and laws governing conflict, they should not blind one to the fundamental challenge: the world today is experiencing unpredictable and accelerating evolution across the entire technological frontier. This evolution is turning the planet, most human institutions, and the human itself into design spaces, with effects that we can only begin to guess at. The military and security challenges this poses are made more daunting because existing models and analyses aren't entirely obsolete, they're just not able to adequately explain new realities. Indeed, as in physics, where quantum mechanics didn't replace Newtonian worldviews and models entirely, but extended the explanatory power of physics into realms where Newtonian concepts failed, so here. Traditional authorities such as Sun Tzu, Carl von Clausewitz, and Antoine-Henri Jomini are not obsolete; rather, the realm of conflict has grown far more complex, with traditional military confrontations becoming a smaller part of a more difficult and multidimensional whole.

The opinions and views expressed in these essays reflect the many discussions I have been lucky enough to enjoy with colleagues and critics in the course of preparing this manuscript, and the articles on which some of the essays are based. I would like particularly to thank Dan Sarewitz, Joel Garreau, and Daniel Rothenberg for their insightful comments. Of course, not all of them agree with me, and the responsibility for errors, faulty analysis, and loose thinking is all mine.

1

ON CONFLICT IN THE 21ST CENTURY

Introduction

Few things have been as consistent throughout human history as violent conflict: violence among families, clans, and tribes; violence among institutions, interest groups, and classes; violence among kingdoms and states; violence among religions and cultures. Scholars debate whether modern levels of violence are higher or in fact lower than historical trends, and if so by what measures. Others argue that war itself, despite being an obvious evil, is in fact good, in that it leads to the creation of stable institutions such as empires or states within which people can prosper far more than in less ordered environments. Such discussions seem beside the point to the many people who see the world today as chaotic and dangerous, increasingly riven by religious and ideological fundamentalism, with the traditional responses of the state, such as deployment of conventional military power, less and less effective—even counterproductive. Indeed, terrorist events call into question one of the most fundamental duties of the state: to protect its citizens.

This chapter will argue, however, that such pessimism is unwarranted. The world today is indeed marked by violent conflict, with even the few efforts to create moderating institutions, such as the European Union, mired in bickering and potentially fragmenting with Great Britain's recent vote to exit the union. Moreover, informal and even much organized violence no longer seems to be even remotely tied to traditional state-to-state conventional military activity, and terrorism in various guises dominates headlines and politics around the world. But fundamentalists, nationalists, anarchists, and revolutionaries have been violent for centuries. Despair, also, is not new; consider, for example, the opening stanza of W. B. Yeats' poem "The Second Coming," written in 1919 as the devastating impact of World War I on the optimism of the Enlightenment became clear:

Turning and turning in the widening gyre
The falcon cannot hear the falconer;
Things fall apart; the centre cannot hold;
Mere anarchy is loosed upon the world,
The blood-dimmed tide is loosed, and everywhere
The ceremony of innocence is drowned;
The best lack all conviction, while the worst
Are full of passionate intensity.

Not only are fear and undue pessimism unwarranted, they are downright dangerous. While it is easy to conflate conflict with violence, that link is neither necessary nor causal; indeed, much of the conflict in the world today is not violent. The U.S. election process, the European Union's struggles with immigration and managing the Euro, the tension between the rising power of China and the United States as existing superpower—all of these situations involve conflict but, to date, very little violence. Today's conflicts, misunderstood and over-simplified, and perhaps even more complex than in the past, can easily slide from words and posturing to violence.

This is not new, as anyone with a passing knowledge of the beginnings of the 20th century's wars will realize. But in a world where accelerating change in many social, cultural, technological, institutional, and economic domains is fragmenting and weakening governance mechanisms, rejecting fear and oversimplification is even more critical, lest they undermine stability and lead to unnecessary violence.

The distinction between violence and conflict leads to a fundamental point: *properly understood and managed conflict may be an important mechanism for reducing actual violence.* Conflict is part of the human condition, and given the current period of rapid and accelerating technological, cultural, and geopolitical change, will not diminish in the foreseeable future. To reject conflict at this point in history is to reject reality. But conflict does not have to be violent, destructive, and mindless. The purpose of studying military and security strategy, doctrine, and conflict management as a new and integrated domain is precisely to shift conflict onto more ethical, hopeful, and beneficial paths. Violence is one way to express conflict; it is not synonymous with conflict.

In particular, there is an urgent need to understand the current characteristics of conflict in broad overview, the better to at least partially steer conflict from unproductive expression—the chaotic and violent state of much of the Middle East and sub-Saharan Africa, for example—to productive. An example of the latter might be African countries competing to design commercial legal systems so that new businesses can be easily started and funded, thus creating jobs and the potential for economic and social prosperity. Letting China's rise to superpower status trigger an unnecessary war with the United States represents a failure to manage conflict, because misses the opportunity to enable that rise so that it is both peaceful and beneficial to both societies. Among other things, however, this would require developing the capabilities and institutional frameworks that enable conflict management.

Wishing for a world without conflict is both impractical and, because it reduces the opportunities to identify and work toward productive, nonviolent outcomes, undesirable. This chapter will instead illuminate some of the dynamics that characterize current global geopolitical conflict, many of which are indeed new, complex, and challenging. The emphasis will be on military strategy and analysis, not because that should be the primary or exclusive way to approach conflict, but because it is both an important perspective and an accessible way into the complex terrain of modern conflict. The chapter will close by suggesting some ways that geopolitical tensions can be managed so as to encourage constructive conflict, and discourage destructive conflict. An important opportunity to redirect at least some conflict towards less violent expression currently exists, and should not be squandered.

And there are plenty of troubled areas around the world where violent conflict could quickly erupt. The Russian conquest of Crimea and invasion of eastern Ukraine in 2014, following on earlier Russian military actions involving Georgia, Transnistria, and other post-Soviet entities, appears to signal a Russian strategy of reasserting regional control, conducted in some cases through direct military action, but in other cases through more subtle means.[1] Jihadist Islam continues to challenge nation-states internally and externally through flexible global networks with no state-based core. Managing the rise of China in relation to the United States, and the evident opportunities for confrontation and misunderstanding in areas such as the South China Sea, raises a number of challenges for both countries, as well as many Asian states. North Korea, the India-Pakistan confrontation, and Iranian military modernization raise threats of regional nuclear conflict.

---

[1] And always carried out with the specter of massive nuclear capabilities in the background.

Each conflict has its own distinct characteristics. The traditional formulations of military competence and national security strategy do not fully capture the challenges that emerge from these characteristics, nor do they provide a comprehensive framework for addressing them. Jihadist Islam will not fall in a traditional military campaign that uses mass forces directed at a decisive point, because there are no decisive points. Russian subversion extended to the Baltic states will be difficult to defeat through deployment of conventional military force. Chinese cyberattacks will not be successfully countered by, in the words of some American officials, "putting a missile down a smokestack." Some areas of the world, especially in the Middle East and Africa, may revert to "neo-medievalism," as state-based models of governance fail and a complex mixture of institutions, religions, clans, and interests form a constantly shifting structure of incoherent governance—"durable disorder," in the words of Sean McFate in his 2014 book, *The Modern Mercenary*.

Two thousand years of strategic thinking are not suddenly invalid. Military action or the threat of such action remains core to managing a complex and confusing world, and traditional state-to-state conflict (e.g., American efforts to manage Iran and North Korean nuclear weapon programs) will remain critical. But the strategic question is no longer simply one of coupling political goals to military means. Rather, effective strategy must define, understand, and successfully manage conflicts that involve not just military and security domains, but economic, institutional, political, technological, social, and cultural dimensions as well.

American Dominance in Conventional Military Domains

At this point in history, the conventional military forces of the United States are stronger than those of any other power. Consider, for example, military budgets. In 2012, total direct global military expenditures were approximately $1,756 billion, of which fully $685 billion was accounted for by the United States alone. In contrast, China spent $166 billion, Russia $91 billion, and the United Kingdom $61 billion. Some countries in less settled regions devote a much greater share of their economies to the military than the major powers: Israel dedicated 7.4% of its gross domestic product (GDP) to defense expenditures, and Saudi Arabia 9.1% of GDP (2012 figures). But no single power spent anywhere near what the United States did in total.[2]

In part, this disparity is a result of the wars of the past century. After World War II, the United States was the only major power not domestically ravaged by war, and the country entered into the Cold War determined to contain and outlast the Soviet Union through, in part, spending on its military. Another reason for the U.S. military's dominance is more recent: as a result of its wars in Afghanistan, Iraq, Bosnia, and elsewhere, the United States currently has a battle-tested military, experienced not just in combat but in equally important domains such as force integration and logistics management.

---

[2] "Country Comparison: Military Expenditures," *The World Factbook*, U.S. Central Intelligence Agency (2013), available online: https://www.cia.gov/library/publications/the-world-factbook/rankorder/2034rank.html; "SIPRI Military Expenditure Database," Stockholm International Peace Research Institute (2013), available online: https://www.sipri.org/databases/milex.

This dominance has had the obvious effect of driving competitors to seek strategies and domains where they can avoid direct confrontation while still asserting their interests. This strategic imperative, combined with accelerating evolution across the entire technological frontier and the increasing democratization of these emerging technologies, leads to "asymmetric warfare" strategies. Such strategies emphasize non-traditional domains of conflict like cyberwarfare, where American dominance is less marked, or the United States is more vulnerable, or both. These strategic and technological imperatives are changing how war and conflict are framed, resulting in a shift from military confrontation to a much broader and complex conflict waged across all domains of civilization, including the economic, cultural, and ideological.

The examples discussed below of the Russian invasion of Crimea, Chinese exploration of "unrestricted warfare," and jihadist Islam provide some insight into what this evolution might look like. They suggest that managing conflict by redirecting it to the extent possible from destructive to constructive modes is both feasible and the only realistic response for reducing the magnitude and consequences of conflict.

To be clear, little of this is new in the sense of being a discontinuous break with the past. What is new is the rate of change; the volume and velocity of information that must be understood and managed; the degree to which emerging technologies are undermining old assumptions, verities, and institutions; and the concomitant increase in complexity and unpredictability that result. Taken together, these forces are altering the world not just in degree, but in kind, and are posing a profound challenge to military and security policies and institutions.

The Russian Invasion of Crimea

Many observers have noted that Russia's 2014 Crimean operation and its subsequent invasion of eastern Ukraine were nontraditional. Some argue, in fact, that no invasion, at least as usually defined, even occurred. Although a number of terms are used to describe Russia's strategy—American officials and analysts from the North Atlantic Treaty Organization (NATO) have called it "hybrid warfare"—Russian military authorities use the term "New Generation Warfare," which is as apt as any other.

An introduction to the thinking behind the strategy was provided in an article by General Valery Gerasimov, Chief of the General Staff of the Russian Federation. He notes that in the 21st century there has been "a tendency toward blurring the lines between the states of war and peace," and that "a perfectly thriving state can, in a matter of months and even days, be transformed into an arena of fierce armed conflict, become a victim of foreign intervention, and sink into a web of chaos, humanitarian catastrophe, and civil war."[3] He continues:

> *The very "rules of war" have changed. The role of nonmilitary means of achieving political and strategic goals has grown, and, in many cases, they have exceeded the power of force of weapons in their effectiveness.*

Although General Gerasimov was writing before the February 2014 Russian invasion of Crimea, his article provides the strategic blueprint for the invasion:

> *The focus of applied methods of conflict has altered in the direction of the broad use of political, economic, informational,*

---

[3] General Valery Gerasimov, "The Value of Science in Prediction," *Voenno-Promyshlennyi Kur'er* [*The Military-Industrial Courier*] (27 Feb. 2013), translated by Rob Coalson and available online: https://inmoscowsshadows.wordpress.com/2014/07/06/the-gerasimov-doctrine-and-russian-non-linear-war/.

humanitarian, and other nonmilitary measures – applied in coordination with the protest potential of the population. All this is supplemented by military means of a concealed character, including carrying out actions of information conflict and the actions of special-operations forces. The open use of forces – often under the guise of peacekeeping and crisis regulation – is resorted to only at a certain stage, primarily for the achievement of final success in the conflict.

The successful Russian use of New Generation Warfare in its Ukrainian invasion is of great interest to those Eastern European countries that fear they may be next. Latvian defense experts, for example, have explored Russian's combination of psychological warfare, political subversion and penetration, intimidation, bribery, Internet and media propaganda, and minimal formal combat personnel in the Crimean and Ukrainian theatres, and how these various factors were effectively integrated into political, psychological, and information strategies. This selection of means was deliberate, as Russia's goal was not traditional conventional military confrontation, or the annihilation of opposing military forces in a climactic event, but rather the destabilization of traditional social and governance institutions and the subversion of any viable opposition.

Indeed, Russia sought to avoid the appearance of traditional military aggression, which might trigger a wider conventional military conflict. For example, Article 5 of the North Atlantic Treaty that established NATO provides that an "armed attack" against one signatory constitutes an attack against all signatories, thus triggering mutual aid provisions. Does the long-term subversion of a state or part of a state using primarily local assets and criminal—rather than military—violence constitute such an "armed attack"? Of course, NATO is quite aware of the implications of New Generation Warfare. The real question is whether the apparent ambiguity of the situation will enable politicians to avoid being called on to honor their NATO treaty obligations to respond.

This form of warfare, which integrates political action, concealed military activities at important leverage points, and sophisticated destabilization initiatives, cannot be considered "new." It is solidly within the Marxist tradition: Karl Marx and Friedrich Engels, in part because of the successful state stifling of revolutionary uprisings in 1848 and later, understood conflict to be diplomatic, psychological, economic, social, and only in its last phases military. The Crimean takeover was not dissimilar to strategies which the Soviet Union and others practiced internally and externally for many years, whether or not they conceived of them as Marxist.[4]

New Generation Warfare is also a comfortable strategy for a state currently led by an individual, Vladimir Putin, whose professional experience was in the espionage and state security apparatus, rather than the military. It is thus not surprising to see the tactics of spycraft arise in a new and effective guise. Whatever one may think of the strategy, it is now a proven and effective use of asymmetric tactics to avoid conventional conflict.

Emerging technologies have altered the context within which New Generation Warfare operates. New information and communication technologies (ICT), and the effect of moving from information scarcity to information overload, significantly change the ways in which information can be deployed and weaponized. Eric Schmidt,

---

[4] In fact, the United States used the method to some extent in supporting right-wing Contras against the socialist Sandinista Junta of National Reconstruction government in Nicaragua from 1979 to the early 1990s. This case may be of particular interest because the Reagan administration covertly continued the program even when the U.S. Congress banned it, so it was subverting not just Nicaraguan, but U.S. governance.

the executive chairman of Alphabet (Google's parent company), once observed that humans now create more information every two days than was created in all of human history up to 2003, and the information that Google processes every 250 days is roughly equivalent to all the words ever spoken by humanity. The specifics may be imprecise, but few disagree with the basic point that the growth in information volume and velocity is accelerating and unprecedented.

Modern search engines give everyone with access to the web the accumulated memory of much of humanity. A few over-the-air television stations have been replaced by YouTube and other video services with rates of information flows that previous generations could not have dreamed of. Society is awash in facts, videos, information, tweets, posts, advertisements, analyses, and blogs. This information tsunami in turn psychologically pressures individuals to retreat to a coherent information structure, supported by filters that limit the information to which individuals are exposed. These filters shape narratives that create what might be called "ring-fenced reality."

Ring-fenced reality depends not on state control of all information sources, but rather on the creation and support of a belief system—a powerful and simple narrative, or what Adolf Hitler called "the Big Lie" in *Mein Kampf*—that can be maintained within a much larger and chaotic information system. Ring-fenced reality is created through adroit manipulation of culture, psychology, beliefs, ideology, perceptions, and opinions using a wide variety of media tools (e.g., comment boards, blogs, websites, traditional print and broadcast media, manufactured news feeds). The creators may be factions or parties seeking political power, firms seeking economic power, communities or non-governmental organizations seeking particular policy outcomes—or, of course, states engaged in conflicts at various levels.

Manipulation of information is not new—it is, after all, the basis of marketing—but its sophisticated use in today's very different information environment is. Controlling public information sources is unnecessary if various audiences can be manipulated so that their psychological and institutional filtering mechanisms only troll for that which supports their pre-existing worldviews. When it comes to information management, then, New Generation Warfare is indeed something new, to the extent its practitioners understand and exploit the modern ICT environment. Moreover, it should be no surprise that potential competitors of a conventional power such as the United States should be interested in the asymmetric opportunities offered by ICT. Such a new and rapidly changing domain is exactly the kind of conflict space where existing, powerful, and successful organizations would find it difficult to adapt smoothly and rapidly.

Chinese Strategy: "Unrestricted Warfare"

Russia is not alone in rethinking its strategy in light of American conventional dominance. Shocked by the success of allied forces in Desert Storm, Chinese strategists Qiao Liang and Wang Xiangsui began thinking along the same lines. Rather than New Generation Warfare, however, the Chinese are developing an overall strategy of "Unrestricted Warfare" based on the perspective that warfare in the 21st century is, and will be, qualitatively different than in the past:

> *[T]here is reason for us to maintain that the financial attack by George Soros on East Asia, the terrorist attack on the U.S. embassy by Usama Bin Laden, the gas attack on the Tokyo subway by the disciples of the Aum Shinri Kyo, and the havoc wreaked by the likes of Morris, Jr. on the Internet, in which the degree of destruction is by no means second to that of a war, represent semi-warfare, quasi-warfare, and*

sub-warfare, that is, the embryonic form of another kind of warfare.[5]

Unrestricted Warfare is closer than New Generational Warfare in moving toward a coherent theory of 21st century conflict because it contemplates the inclusion of all dimensions of a civilization in a deliberate, strategically integrated process of long-term, intentional, coordinated conflict, one aspect of which may or may not be conventional combat.

Again, this is a case of something old, something new. The Cold War, for example, involved competition not just militarily, but in culture, economics, and many other domains. In the case of Unrestricted Warfare, emerging technologies such as ICT, robotics, big data analytics, and artificial intelligence (AI) have dramatically increased conflict's complexity and unpredictability, not to mention the opportunity for strategic disruption. As Qiao and Wang note, "As we see it, a single man-made stock-market crash, a single computer virus invasion, or a single rumor or scandal that results in a fluctuation in the enemy country's exchange rates or exposes the leaders of an enemy country on the Internet, all can be included in the ranks of new-concept weapons."[6]

The implications of such a perspective for traditional military thinking are profound:

---

[5] Qiao Liang and Wang Xiangsui, *Unrestricted Warfare*, U.S. Central Intelligence Agency translation (Beijing, China: People's Liberation Army Literature and Arts Publishing House, 1999), p. 3. The CIA version is available on the web at http://www.c4i.org/unrestricted.pdf, and is highly preferable to other versions that distort the original content, such as the book version entitled *Unrestricted Warfare: China's Master Plan to Destroy America,* which on its surface takes a rational analysis of the United States from a Chinese perspective, and tries to sensationalize it. Citations in this chapter will be to the CIA version.

[6] Ibid., 13-14.

> *Faced with warfare in the broad sense that will unfold on a borderless battlefield, it is no longer possible to rely on military forces and weapons alone to achieve national security in the larger strategic sense.... Obviously, warfare is in the process of transcending the domains of soldiers, military units, and military affairs, and is increasingly becoming a matter for politicians, scientists and even bankers.... Think about the Lockerbie air disaster. Think about the two bombs in Nairobi and Dar es Salaam. Then think about the financial crisis in East Asia.... This is warfare in the age of globalization.... If those such as Morris, bin Laden, and Soros can be considered soldiers in the wars of tomorrow, then who isn't a soldier? If the likes of Powell, Schwarzkopf, Dayan, and Sharon can be considered politicians in uniform, then who isn't a politician?*[7]

The inclusion of Soros as a "soldier in the wars of tomorrow" is interesting because it assumes that financial systems are weapons and, by extension, legitimate targets. Qiao and Wang comment that "financial war has become a 'hyperstrategic' weapon that is attracting the attention of the world. This is because financial war is easily manipulated and allows for concealed actions, and is also highly destructive."[8] This perspective provides the potential for misunderstanding between westerners, who perceive market and financial activities as clearly part of a civil sphere, and the Chinese, who will be inclined to see financial actions by competitor nations such as the United States as strategic moves in a multidimensional conflict.

Conflict at the level of cultures and civilizations does not necessarily imply greater violence, or an increase in traditional military operations. Since the ascending power (in this case China) will try to avoid attack at the point of their opponent's strength (American conventional military strength), asymmetric conflict involving ICT and similar

---

[7] Ibid., 118-119.

[8] Ibid., 27.

mechanisms, and economic and cultural initiatives, is likely to be the order of the day—although the possibility always exists that bluster, or operational mistakes, trigger armed conflict.

Furthermore, there are indications that the Chinese perceive future conflict to be less lethal. As Qiao and Wang observe: "mankind is... beginning to learn to control the lethal power that it already has but which is increasingly excessive... kinder weapons represent the latest conscious choice of mankind... thereby giving warfare an unprecedented kind-hearted hue."[9] Whether this hopeful vision holds in the real world of maneuvers in the South China Sea and elsewhere, of course, is a different question. The 2015 U.S. National Military Strategy notes that "We support China's rise," but "its claims to nearly the entire South China Sea are inconsistent with international law."

Unrestricted Warfare presents an interesting illustration of the possibilities inherent in a more sophisticated understanding of conflict. Contrary to what initial impressions might indicate, deliberate, strategic inclusion of virtually all domains of a culture in explicit conflict with its adversaries may actually enable a reduction of violence. But leaving that possibility to chance would be unwise.

Jihadist Islam: The Non-State Actor

Despite New Generation and Unrestricted Warfare strategies, both Russia and China as successful states are relatively comfortable within, and able to assert their interests through, a state-based, or "Westphalian," world order. Jihadist Islam, on the other hand, is an example of a powerful actor taking a radically different perspective. As

---

[9] Ibid., 15.

former U.S. Secretary of State Henry Kissinger notes in his recent book, *World Order*:

> *This body of [Islamic] thought represents an almost total inversion of Westphalian world order. In the purist version of Islamism, the state cannot be the point of departure for an international system because states are secular, hence illegitimate; at best they may achieve a kind of provisional status en route to a religious entity on a larger scale. Noninterference in other countries' domestic affairs cannot serve as a governing principle, because national loyalties represent deviations from the true faith.... Purity, not stability, is the guiding principle of this conception of world order.*[10]

In this case, one does not find a sophisticated redefinition of conflict in terms of asymmetric power and opportunity, based on traditional geopolitics and military strategy. Instead, there is a rejection of the entire framework of modernity.[11] Truth and knowledge, for example, are not understood as products of science and applied rationality, but rather as products of faith. Religious systems, not state or secular systems, are the sources of social, legal, and institutional organization. Universalist values do not arise from secular constructs such as "human rights," but from the underlying belief system.

Absolutist perspectives rejecting the existing international framework of institutions, agreements, and norms are not common, but neither are they limited to jihadist Islam. Indeed, many fundamentalist movements, including not just other religions but also powerful ideologies such as environmentalism, elevate their belief systems above the pluralistic framework implicit in the international system, for the most part without violence. Fur-

---

[10] Henry Kissinger, *World Order* (New York, NY: Penguin Press, 2014), p. 122.

[11] Kissinger, I'm obliged to note, is talking about a particular strain of extremely conservative, fundamentalist Islam, not the entire religion or culture.

thermore, although their violent acts may capture headlines, the operational challenge terrorist groups pose seldom rises to the level of conventional warfare (ISIS in the Middle East, Nigeria's Boko Haram, and Al Shabaab in the Horn of Africa being notable exceptions). Most fundamentalist conflict generated by non-state actors so far has been small-scale terrorist attacks or criminal incidents involving civilians; where large-scale maneuver has been deployed, as with ISIS and Boko Haram, it has usually occurred where state control is weak or non-existent.

In some cases, conventional military tactics such as air strikes and special operations activities may be appropriate responses to the actions of these groups. Complacency is generally inappropriate given the potential for terrorist deployment of weapons of mass destruction. But broad-based rejections of modernity cannot be adequately countered by military means alone. Any adequate response must recognize that such fundamental challenges to world order arise from deep and multidimensional causes, and thus require an appropriately multidimensional conflict strategy.

For example, one stressor for many individuals and institutions is accelerating technological, social, and cultural change, which undermines many strong and heretofore unquestioned cultural beliefs and social practices. An understandable response is a retreat to faith on the part of those who are unable to keep pace with, or accept the changes inherent in, such a world. The accelerating complexity of an increasingly multicultural world may have the same effect, reinforcing the value of mythic cultural stereotypes and narratives of past "golden ages" as refuges. This pattern may be evident in phenomena such as the Tea Party in the United States, the return to tsarist patterns in Russia, and perhaps most obviously in the effort of ISIS to re-establish a caliphate in the Middle East. If the immediate military threat of ISIS can be managed through military responses, the reasons ISIS is there in the first place—the retreat from the challenges and changes inherent in modernity—cannot.

From Conventional Military Engagement to Asymmetric Warfare to Multidimensional Conflict

Three fundamental principles of strategy help explain the new aspects of geopolitical conflict described above. First, those opposed to the status quo should to the extent possible develop tactics and strategies that reflect their relative strengths, and avoid the strengths of their enemies. Second, in the real world, the adversary always gets a vote. The Prussian general and theorist Carl von Clausewitz expanded this idea to include not just the adversary but the context in which conflict takes place. He introduced the concept of friction: the accumulation of inevitable small events and challenges in complex operations which make any strategy, no matter how sophisticated, inadequate unless constantly revised and tested in real time under actual conditions.[12]

Third, major technological changes often incentivize asymmetric behavior, which in turn fosters innovation in military and security technologies and strategies. This dynamic can be seen in the examples above. Russia repurposes the idea of "Novorossiya"—a historical term for the region north of the Black Sea that was traditionally Russian territory—as part of its Ukrainian invasion strategy through manipulation of modern ICT environments. China challenges the West generally, and the United States

[12] "Everything in war is very simple," he wrote, "but the simplest thing is difficult. The difficulties accumulate and end by producing a kind of friction that is inconceivable unless one has experienced war." Carl von Clausewitz, *On War*, ed. and trans. by Michael Howard and Peter Paret (Princeton, NJ: Princeton University Press, [1832] 1976).

specifically, through deployment of Internet tools intended to steal commercially and militarily valuable information. ISIS uses social networking to distribute a potent vision of identity to lost and alienated Muslim adolescents.

The impact of technology on conflict can be seen throughout history. In the 1430s, the French modernization of cannon in Europe made existing fortifications obsolete. Within a decade the Italians had figured out how to redesign fortresses with thicker sloped walls and protective fields of fire to be able to counter such assaults (the new style was dubbed *trace italienne*). Nuclear weapons introduced a technology that, after a period of technological and strategic evolution, was generally considered by the states involved to be so terrible as to be essentially unusable in actual combat. The result was not just a technological response but a strategic innovation, the doctrine of Mutually Assured Destruction (MAD). A final example can be found after World War II, when peoples seeking independence from the dominant conventional forces of colonial governments developed the strategies and tactics of guerilla warfare, in turn spawning the study of counterinsurgency strategies.

Key Themes of the Re-Characterization of War

Given the American dominance in conventional military technologies, potential adversaries of the United States have been pushed toward a deep reframing of the idea of conflict, because all the easy or incremental routes for seeking asymmetrical balance of power are inadequate. Once this is understood, a few basic rules of this reframing emerge:

1. **Reconceptualize the role and purpose of military action in conflict.** Both the Russian and the Chinese strategies identify military action as the last step in a

particular conflict, to be taken only when victory has already been assured through other means. In a clear break from traditional military strategy, the point of conflict is not mere victory in a particular battle, but rather success in the long-term competition between cultures, a process that is viewed as only occasionally requiring conventional conflict.

2. **Expand the definition of conflict to include multiple domains.** Because traditional conventional military action has become too costly and risky, domains such as civilian cyber realms are more favorable for conflict. This can lead to the utilization of non-traditional opportunities: some jihadist groups, for example, have identified the alienation of adolescent males as a domain where social networking tools and relatively sophisticated psychology can be fairly effective recruitment tools.

3. **Reject traditional ideas regarding what constitutes a "successful action."** Conventional military dominance defines military success in terms of definitive battles or campaigns. Russia's Crimea invasion was, however, a success for the Kremlin and New Generation Warfare even without a definitive victory, and the doctrine of Unrestricted Warfare supports the apparently successful and certainly notable level of Chinese cyber-activity.

4. **Design and operate conflict activities below levels that may trigger conventional military responses.** Russia's attack on Ukraine was riskier than China's cyber-campaign against American military contractors and corporations generally because it did involve elements of conventional force. Use of New Generation Warfare techniques, however, including both direct threats to European energy supplies and other economic interests, and more subtle exploitation of differences between American and European allies, reduced

the possibility of any united military response. Russia's weaponization of Edward Snowden's leaks about U.S. surveillance to drive a wedge between the Americans and the Germans was particularly adept.

5. **Integrate conflict activities across society as a whole, rather than isolate them in a dedicated military organization.** As Qiao and Wang observe, "there is no longer any distinction between what is or is not the battlefield... social spaces such as the military, politics, economics, culture, and the psyche are also battlefields."[13]

Multidimensional Conflict Competence

Innovation in asymmetric competition and conflict, combined with rapid evolution across virtually the entire technological frontier, has created non-traditional geopolitical dynamics. These dynamics may favor competitors such as China and Russia (the effect on non-state and private actors is less coherent). Both of these countries tend to have less stringent demarcations among their industrial, civilian, and military domains, and less of an emphasis on the rule of law, than in the United States.

For example, American tradition beginning with George Washington has reinforced a divide between the military and civilian spheres.[14] This is an American strength that makes U.S. social, economic, and institutional structures innovative and adaptable; but it is also a potential barrier to integrating conventional military

[13] Qiao and Wang, op. cit., p. 110.

[14] Article I, Section 8, Clause 11 of the U.S. Constitution gives Congress the power to declare war; Clause 14 gives Congress the power to "make Rules for the Government and Regulation of the land and naval Forces..." Article II, Section 2, Clause 1 makes the President, not the highest military officer, the "Commander in Chief of the Army and Navy of the United States..."

strength into other domains of conflict. On the other hand, more authoritarian states' ability to integrate across their systems more effectively may encourage costly "group think" and less agility when faced with unpredictable and rapidly evolving threats.

Another aspect of governance in the United States and some European countries is the separation of the private from the public sectors, a sometimes-hazy border between the management and direction of commercial activities and the government. Of course there are regulatory structures, joint activities, lobbying, corruption, and the like, but American firms largely plot their own course, especially compared to firms in Russia, China, and other countries characterized by a more state-based form of capitalism.

The political environment in other countries might also be more favorable to new forms of conflict. Some adversaries may be more adept at engaging in long-term conflicts than the Americans, an advantage for strategies built on incremental gains in diverse domains. The Chinese, for example, are known for taking a long-range view, especially on questions involving their evolution towards superpower status, while non-state actors motivated by religious perspectives may be similarly capable of maintaining their institutional coherence over long periods of time.

In short, the evolution away from traditional state-to-state military conflict and toward new varieties of multi-dimensional conflict across military, security, economic, civil, and social domains may challenge conventional powers. These conventional powers, such as the United States, are characterized by a) rule of law; b) constitutional and legal boundaries between the military and civilian sectors; and c) effective separation of the commercial and governmental spheres. These characteristics may become a weakness for conventional powers as non-state political, religious, and commercial entities evolve in size and complexity to challenge traditional state dominance of large-scale conflict.

But the picture is not completely one-sided. The United States does not lack for its own comparative strengths, particularly in the realm of "soft power," which comes to the fore as conflict is redefined away from conventional military confrontations and towards different flavors of multidimensional conflict. The firms that create American popular culture, packaged and sold through music, media, and commercial brands such as Apple, Coca-Cola, Levi's, and McDonald's, are globally dominant, economically formidable, and an important part of American power projection.

The American higher education system remains among the world's best, although constraints on immigration, rapid changes in educational technology, and foreign competition from countries such as Australia and the United Kingdom are eating away at its dominance. The American economic framework supports and rewards innovation and entrepreneurship, and provides venture capital and managerial support for start-up firms (despite some remarkably shortsighted policies in areas such as immigration). Accordingly, the country remains one of the most attractive environments for innovators and creative individuals from around the world. Silicon Valley, for instance, is a global village that would be difficult to replicate in another culture.

The rapidly accelerating complexity of geopolitical conflict has several implications. Perhaps most important is understanding that, because of this complexity and the different domains involved in modern conflict, centralized, top-down control is simply dysfunctional. Any single perspective, no matter how enlightened, is likely to be at best partial and arbitrary. In the case of the United States, its soft power dominance arising from its open society is one of the country's great strengths—and a strength that could be jeopardized by inappropriate, if tempting, efforts to control and direct the system (through, for example, any effort to try to militarize the soft power of American brands and entertainment).

In fact, all three of the major competing nations discussed here—Russia, China, and the United States—are attempting to find different ways of balancing centralized direction with the agility, responsiveness, and adaptability that derive from pluralism. Russia tends to emphasize centralized control; China and the United States are experimenting in different ways with a balance that encourages more diversity. In addition to being experiments in governance, these efforts may signal the beginning of a new phase in the competition among these three claimants to global power status.

Conclusion

At the beginning of this chapter, I pointed out that encouraging constructive conflict rather than destructive conflict is an important reason to develop a better understanding of conflict itself, particularly in light of growing geopolitical complexity, increased uncertainty, and rapid change across the entire technological frontier. But that unpredictability, uncertainty, and complexity also make it difficult to offer specific paths forward. Accordingly, without any pretense at comprehensiveness, I conclude by noting several potential elements that a successful program to manage modern conflict, in the United States or elsewhere, might include:

1. **Explicitly recognizing the existence of the new doctrines, strategies, and tactics of multidimensional conflict, along with their implications.** This may require revisions to existing institutions. For example, are current legal structures intended to manage kinetic

conflict, such as the Laws of Armed Conflict and institutions such as the North Atlantic Treaty Organization, still adequate, and if so under what circumstances? What additional legal structures should be contemplated? How can existing laws governing conflict, especially violent conflict, be modernized as emerging technologies and geopolitical developments undermine the assumptions on which they are based?

2. **Developing a sophisticated coordination process—but not "control," which is not feasible or effective with complex systems—across the many domains of society that are implicated in multidimensional conflict.** The process's goals should be 1) to achieve a distributed but coordinated information structure that enables identification of and response to threats and opportunities; and 2) to encourage constructive rather than destructive conflict. Conflict is inevitable but need not always be violent, and can to some extent be managed; violence, at least in some cases, can be decoupled from conflict and reduced where it must occur.

3. **Remembering that conflict and competition in nonmilitary domains augments, but does not replace, existing military, terrorist, and criminal challenges.** A strong strategy should enable identification, development, and deployment of U.S. military and nonmilitary conflict assets in ways that facilitate the goals of point 2 above.

4. **Developing a theory and strategy of integrated multidimensional conflict.** Traditional military dominance, while necessary and appropriate in some cases, is unlikely to be systemically effective given the complexity of the challenge. This suggests that new institutional structures will be required. Pluralism is a valuable problem-solving mechanism in complex systems, and may thus be an important asset in high-level, long-term, multidimensional conflict environments.

5. **Building the institutional ability to extend operations over long time frames and to protect conflict capabilities from the vicissitudes of political turmoil.** The rise of dangerously simplistic right- and left-wing populism in the United States, Europe, and elsewhere indicates that this will be no simple task. This also raises questions regarding the ability of militaries to compensate for growing domestic political weaknesses.

6. **Recognizing that traditional military assets remain critical to maintaining great-power status, but that aggressive actions are problematic in the context of multidimensional conflict.** This point is most apparent in in counterinsurgency operations where the goal is to "win hearts and minds," but is also valid at the level of cultural competition among major adversaries. China, for example, has made life much harder for itself by its aggressive actions in the South China Sea, which have alienated potential regional allies. Major powers that are able to provide credible assurances to other countries—and even to their own populations—of the benefits to all of their superpower status are likely to enhance their soft power, and thus their overall conflict capabilities.

The world cannot escape conflict, but it can limit unnecessary violence and destruction that such conflict might otherwise imply. Doing so will require more imagination, creativity, and institutional agility than currently exists, but is well worth the effort.

2

ETHICS OF EMERGING MILITARY TECHNOLOGIES

Introduction

Discussions regarding military and security technologies often focus on operational dimensions such as performance, tactical implications, and potential defense strategies. Hypothetical social, cultural, and policy implications of technologies are also frequently debated, often heatedly, both because the technology itself can be contentious—think of drones, for example—and because military and security technologies are often capable of inflicting damage and death. Lost in the technical details of the former discussion and the hyperbole of the latter debates is the need for serious appraisal of the ethical implications of military and security technologies.

The link between technology and military and security domains has existed throughout history, but it is particularly important today. Accelerating evolution of technology across its entire frontier is combining with the ability of states and violent non-state actors to use civilian technology to inflict damage on adversaries. Terrorists use airplanes and mobile phones as bombs. Teams of hackers attack perceived enemies and sometimes augment traditional combat initiatives through disinformation campaigns intent on sowing confusion and misdirection. Terrorist groups use social media platforms to recruit new members and plan attacks.

At the other end of the military-civilian spectrum, technologies developed specifically for military or security applications can have very different effects when they diffuse throughout civil society. Hummingbird drones, for example, are obviously useful in combat, especially in counterinsurgency environments where collateral damage must be minimal, but they have far different implications when adopted by divorce lawyers, political parties, and partisan news organizations.

The importance of emerging technologies in the conflict sphere is amplified by a couple of characteristics of technology systems. First, technologies are not just artifacts: they are social, cultural, and economic phenomena. A powerful technology cannot be deployed for a single purpose, even in the military or security domain; it is integrated into society, thereby changing it. Technologies that may be developed for one military purpose—say, to provide a "smart" prosthesis to a wounded soldier—can be, and often are, turned to other purposes in civil society—to augment existing human capabilities, for example.

Second, technologies of all kinds are being democratized. Many technologies of violence and mass destruction are available to non-state actors, some of whom are highly ideological and violent. In the past, the most advanced military and security technologies tended to rest with the most powerful states, which during the Cold War, for example, were the Soviet Union, the United States, and their principle allies. But now with the rise of asymmetric warfare and the difficulty of keeping technical secrets in the era of Internet espionage, that is no longer the case.

Understanding and ethically managing emerging technologies, particularly in the military and security domains, has always been important. New technologies can provide military advantage, which is why technological evolution and military and security organizations have been tightly coupled throughout history. But technology is also a central link between military and civil society, sometimes changing both in fundamental, far-reaching ways.

The evolution of corned gunpowder in Europe in the late 14th century is a good example. Gunpowder, a combination of sulfur, charcoal, and potassium nitrate, was known to the Chinese by the 9th century, but the ingredients tended to separate out over time, thus reducing its effectiveness. Early mixtures, while they burned impressively, had difficulty providing the rapid energy release necessary for use in weapons. Corned gunpowder, a European improvement, entailed mixing the materials when they were wet and letting them dry in small pellets, which among other effects resulted in a substantial increase in explosive power per unit weight.

This new technology, combined with complementary advances in metallurgy and other sciences, created a much more complex military environment. On the individual level, the development of more portable and powerful guns shifted advantage away from elite mounted knights to infantry. On the institutional level, because a gunpowder army was more expensive than earlier feudal armies and required a more substantial logistics system, gunpowder technology required economies of scale in military activities. Kingdoms, and then nation-states, rather than feudal manors, were best equipped to handle the new expense and complexity of gunpowder armies.[1] Ad-

---

[1] Historians have coined the term "Gunpowder Empire" to refer to empires such as the Turkish Ottoman Empire, the Safavid Empire of Persia, the Mughal Empire in India, and Spain and the Spanish New World, which arose in part as the impacts of gunpowder technology played out over time.

ditionally, the integration of advances in ship design, cannon construction, and gunpowder created the integrated technology platform that enabled European navies to outcompete all others, leading eventually to a world where the British Royal Navy dominated the oceans from the 17th to the 20th century.

Providing a coherent introduction to the applied ethical issues raised by emerging military and security technologies is also difficult in part because conflict has been such a constant in human history. People have been thinking about the ethics of it for millennia. Virtually every major religious text, from the Bible to the Koran to the Tao Te Ching, has touched upon the morality of war. Strategic and historical treatments of war, from Sun Tzu's *The Art of War* (ca. 512 B.C.) and Thucydides' *History of the Peloponnesian War* (ca. 400 B.C.) to Alfred Thayer Mahan's *The Influence of Sea Power Upon History* (1890) to Vasily Sokolovsky's *Military Strategy* (1962), necessarily engage the complex integration of technology, ethics, values, cultural norms, and behavior during military activities.

An example of this, still used as a touchstone for the exploration of deep ethical issues, is the Athenian treatment of the neutral island of Melos in 416 B.C., as described in Thucydides. The Athenians, dependent on and dominant in naval technology, gave the Melians the choice of surrender and tribute, or death. The Melians argued that as a weak, open, peaceful society, justice demanded that they be spared. The Athenians took a "might is right" position: "Of the gods we believe, and of men we know, that by a necessary law of their nature they [the Athenians] rule wherever they can."

In the event, Melos refused, and Athens destroyed their city, killing all the men and enslaving all the women and children. The context is complicated (there is, for example, some evidence that Melos was helping Sparta, Athens' enemy), but the Melian dialog, and the brutality on the part of a society considered the model of civilization, continues to be a fertile source of analysis and discussion. Less frequently noted is how profoundly technology shaped the situation: the Athenians were in a dominant position in the first place in large part due to their mastery of the trireme ship technology, including the tactics of using massed ships effectively and the wealth that their control of the oceans provided.

In our own time, the complexity of conflict itself in the modern age is daunting, and becoming more so due to technological change. Leaving aside technical issues of when a conflict constitutes a "war" under domestic or international law, it was easier in past eras to know when one was being attacked. Conflict was a physical reality, albeit often cloaked in cultural or social patterns and rituals that were and are frequently quite complex (tribal communities often "fought" using highly stylized ritual; even today, the process of declaring war is a form of ritual intended to justify subsequent combat). Now, in an age of cyberconflict, the lines are unclear. Is putting a back door or logic bomb in a software system the equivalent of a kinetic attack? And if it is, given the anonymity of the Internet, against whom does one react?

Since the 1648 Peace of Westphalia, which ended Europe's Thirty Years' War and laid the foundations of the modern nation-state, countries have been the relevant actors under international law and policy, and the entities responsible for initiating, sustaining, and resolving conflict. The United Nations and many other international organizations reflect that assumption. But today's conflict zones often include not just national militaries, governed by prevailing international law, but also non-state actors that may be just as powerful as states, private entities that may be more or less under the contractual control of national governments, independent private actors, and espionage organizations that may be operating under different rules and laws than anyone else.

The lines between war and peace, civilians and soldiers, and civilian and military activities, are increasingly fuzzy. Technological advances enable the blurring of these lines when it is in an actor's interest to do so. Asymmetric strategies such as Russia's New Generation Warfare and China's Unrestricted Warfare, are attractive options against a dominant conventional power such as the United States and are easier to employ than ever. Security in a world dominated by nation-states usually involved identifiable actors and surrogates. Security in today's world, where religious and ideological frameworks support the growth of non-state and informal actors, is a much more complex question. The technologies with some effectiveness in providing security in such an environment, such as big data mining and analytics, may threaten other values, such as privacy, freedom of expression, and freedom of religion. The controversy over the U.S. National Security Agency's surveillance operations, as disclosed by Edward Snowden in 2013, exemplifies such a dialog. Whether and how to deploy technologies then becomes a question of balancing these values, and different cultures will come to different conclusions about what is appropriate.

These considerations intersect with legal structures and norms. This is especially true where war and conflict are involved; a large body of international practice and law has been developed over millennia, with input from many cultures. Many people have heard of the Geneva Convention, for example, and most people who have served in the armed forces probably received some training in the laws of armed conflict. Behind these formulations is a philosophical and ethical line of inquiry going back into antiquity that is referred to as "just war theory." The historical dialog of applied ethics, such as when conflict can be rightfully initiated, how it should be conducted once it is initiated, and how conflict should be terminated, has engaged many thinkers from many different cultures over the ages, up to and including today.

Underlying this legal discourse and its application are supportive cultural norms. These norms, in turn, are underpinned by deep assumptions concerning conflict and what constitutes an acceptable world order. Technological and geopolitical evolution may be undermining these assumptions potentially exposing the more explicit legal and behavioral frameworks erected on them as fragile, and perhaps even unrealistic or dysfunctional. Recognizing that ethical ideas about conflict, human rights, democracy, etc. are not universal is necessary when seeking to minimize violent conflict among actors who may possess very different histories, perspectives, and norms.

Why Applied Ethics?

Theologians and philosophers have explored the large and abstract questions of morality and ethical systems for millennia; modern philosophy includes the field of "moral philosophy," and, more recently, "practical ethics." But in the context of military and security technologies, the complex ethical, social, cultural, and practical issues are beyond the remit of any single discipline. Thus the importance of "applied ethics," a discourse that encourages its practitioners to engage not just in philosophic speculation, but also in public debate and dialog, and to reflect pragmatically on the political, cultural, and industrial systems within which ethical problems arise.

Another difference between applied ethics and more theoretical approaches is where the discourse is housed and where it is focused. The theoretical perspectives are academic discourses, housed in specific disciplines, usually philosophy. Applied ethics, on the other hand, may be informed by academic study, but it is housed in the real world of the practitioner: the professional engineer, lawyer, and military officer, and the institutions of economic, military and political power. The applied ethics discourse emphasizes action in the world rather than a more removed and contemplative perspective. Good argumentation and reason are still critical, but the thoughtful military officer is as likely to present useful perspectives on the applied ethics of military and security technologies as the academic philosopher—perhaps even more likely, since the officer may well have relevant experience with the technologies and their institutional context that the philosopher lacks.

In the United States, a number of trends, such as moving from a conscription to a volunteer military and an increasing geographical segregation by political preference, are converging to separate civilian and military perspectives. (Not being personally acquainted with someone who has served in the military, for example, is possible today in a way that it was not for many decades after World War II.) As former U.S. Secretary of Defense Robert Gates noted in a 2011 speech at West Point,

> *An all-volunteer military is to a large degree self-selecting. In this country, that propensity to serve is most pronounced in the South and the Mountain West, and in rural areas and small towns nationwide—a propensity that well exceeds these communities' portion of the population as a whole. Concurrently, the percentage of the force from the Northeast, the West Coast, and major cities continues to decline.... In addition, global basing changes in recent years have moved a significant percentage of the Army to posts in just five states: Texas, Washington, Georgia, Kentucky, and North Carolina. For otherwise rational environmental and budgetary reasons, many military facilities in the Northeast and on the West Coast have been shut down, leaving a void of relationships and understanding in their wake.*

The same pattern of increasing separation between the military and civilians—and especially civilian academics—seems to be occurring in many countries. This, along with the fact that most philosophers have little to no understanding of technological systems, greatly complicates the philosophical project of studying the applied ethics of emerging military and security technologies. Those who have used technology in military or security environments are more likely to have realistic appraisals of the strengths and weaknesses of the technological system, while those who are inexperienced seem to be more drawn to utopian or dystopian scenarios.

Emerging Military and Security Technologies

The key points about emerging military and security technologies are simple. Technology is evolving rapidly and unpredictably across its entire frontier (even if most discussions only involve a single technology or a related group of technologies—drones, cyber-weapons, or non-lethal ray weapons, for example). Technologies are not simply things: they are implicated in social change. Throughout history, any technological system of any power whatsoever, from the wheel to gunpowder weapons to railroads to the Internet, has destabilized and reshaped the economic, political, cultural, and institutional environment within which it develops.

Past waves of technological innovation have typically revolved around one core system, such as electrification, railroads, or flight. Today, however, technological evolution is occurring across all technological systems, driven by accelerating evolution in at least five core technologies: nanotechnology, biotechnology, information and communication technology (ICT), robotics, and applied cognitive science. Each of these is not just powerful in itself, but an enabling technology that supports unpredictable innovation in many separate domains. For example, nanotechnology enables more powerful ICT, which in turn supports biological printing capability, which will likely lead to the development of 3-D printed replacement organs and maybe factory-grown meat.

Two characteristics of technological change introduce additional complexity. First, technological change is accelerating at historically unprecedented rates. Second, the impacts and processes associated with technological change are unpredictable and nonlinear. These characteristics have geopolitical and security implications, and although they have not been traditionally considered as part of the military and security domains, that may be changing with the concept of unrestricted warfare.

One way these technologies could filter into the military space is through human enhancement. The human is now a design space in ways that it has never been before. While people have always enhanced themselves, the direct interventions that are possible today are far more powerful. Many students, for example, now use widely available drugs that improve concentration and reduce the need for sleep. Vaccines enhance immune systems and powerful prosthetics can integrate human and machine at the tissue level. Scenarios for future human design include development of computer-to-brain interfaces that may enable telepathic technologies and remote operation of robots directly wired into human brains. Even today, the military is a leader in "augmented cognition," or "augcog," where awareness does not occur at the individual level but emerges from integrated techno-human networks.[2]

---

[2] One example of this is the XDATA program, run by the U.S. Defense Advanced Research Projects Agency (DARPA), which funds research on integrated techno-human systems. XDATA can process the information streams generated in modern com-

Some medical researchers claim that the first people who will have a lifespan of at least 150 years, with high quality of life throughout, have already been born in developed countries. As with all claims of coming technology, such assertions are best viewed as scenarios rather than as predictions. Nonetheless, many of the technologies that would support radical life extension are being developed in part by military organizations. The United States, for example, supports a wide variety of research intended to create "super soldiers," some of which might also improve health and—potentially—extend life spans for civilians.

Not only does this illustrate the important point that technologies developed for specific military applications may have wildly unpredictable effects, but it also reveals weaknesses in the institutional and disciplinary tools that might allow their reasoned evaluation. Think for a moment about what radical life extension might mean, for population control, for environmental systems, for pensions and work patterns, for inter-generational relationships. Possibilities proliferate to the point of incoherence. Right now there are no good ways to perceive, much less understand or manage, the challenges of what it means for the human to be a design space.

The example of human enhancement also makes plain how impossible demarcating military from civilian technology can be. Military or security technology used to be defined fairly clearly as something that projected, or protected against, organized physical attack. But any powerful technology can be weaponized: big data analysis powers the security work of the National Security Agency, airplanes are flown into buildings, cyber networks enable attacks on civilian Internet infrastructure. The protean

---

bat far better than either human analysts or computers operating alone.

nature of technology is particularly complex when doctrines and strategy of powers like China and Russia explicitly extend conflict across all civilian and military systems.

Military and security technologies deserve ethical analysis in part because of the profound geopolitical dimensions of these technologies, examples of which can be glimpsed in the discussion above. Stirrups, the composite bow, corned gunpowder, the machine gun, and nuclear weapons don't just change the military balance of power, but contribute over time to the patterns of cultural dominance and failure that constitute history. Technologies like the machine gun enabled European imperial powers to dominate far larger but less well-equipped and trained indigenous military forces, as suggested by the famous doggerel by the British writer and poet Hilaire Belloc: "Whatever happens, we have got / The Maxim gun, and they have not."

But the picture is not that simple. Even though the advantages offered by technological advances, such as corned gunpowder, helped the new technologies rapidly diffuse (and spurred newer technologies in response), the inherent conservatism of military organizations and the high costs of experimentation in conflict environments should anything go wrong can slow technological evolution.

This is especially true where technologies are dual-use, and ethical concerns arise not from the military but from the social domain. Railroads, a potent technology usually not thought of in military terms, provide an interesting example. The strong railroad infrastructure of the North in the Civil War contributed to victory in that conflict, not just because of the obvious advantage rapid rail transport provided in terms of troop movements and logistics, but more generally because the growing rail network constituted a critical infrastructure supporting the North's greater productivity and industrial efficiency.

The Confederacy was at a disadvantage in railroad infrastructure (the North had a track density—track laid per unit land area—that was three times that of the agrarian South) not only because it lagged behind the North technologically. The most significant barriers to rail development in the South were cultural and political. States' rights philosophies meant that Southern railroads were generally not permitted to integrate with each other across state lines, and a fear that rail travel could undermine the agrarian slavery economy led to distrust of the technology. (This pattern repeated in the Austro-Hungarian Empire, Russia, and even France in the years before the First World War; all worried about the social stability of their large and somewhat archaic agrarian systems).

Technologies regarded as purely civilian (or, for that matter, purely military) are in fact generally dual-use. What was true of railroads is also true of modern highway networks, airplanes, robots, and cognitive pharmaceuticals. It is especially true with the Internet and ICT capabilities generally. When countries adopt unrestricted warfare strategies, they use the tools of asymmetric warfare, especially ICT options, which allow them to wear down otherwise dominant opponents. Analysts such as Richard Clarke, a former cybersecurity and cyberterrorism advisor for the White House, have suggested that a "death of a thousand cuts" strategy is the optimal way for adversaries such as China to attack the United States, especially given the unclear status of cyberconflict options under the prevailing laws of war. With such strategies, the line between military and civilian domains becomes at best fuzzy and, depending on circumstances, essentially irrelevant.

Many Americans might consider the "death of a thousand cuts" strategy of degrading civilian ICT systems to be unethical. The applied ethicist must be careful of importing a bias towards existing power structures and concomitant institutional practices. No doubt the British felt that some of the guerilla tactics employed by American colonists were unethical and improper,[3] and Japanese samurai felt that gunpowder weapons were unethical. In post-colonial conflicts, terrorism and insurgency tactics aimed at civilian targets came to be seen as ethical by some, and torture and oppression on the part of counter-insurgency forces to be an ethical response by others. New technologies and strategies mean changes in patterns of conflict, and these may alter what is ethically acceptable.

Applied Ethics and the Laws of War

To understand and assess the applied ethical implications of military and security technologies, it is helpful to start with a brief introduction to the applicable frameworks, formal and informal, which have developed over many centuries around the complex activity of human conflict. While applied ethics are not the same as law, legal structures are an important consideration when evaluating the ethical implications of military and security technologies, and a reasonable guide to the underlying—and often somewhat inchoate—norms.

Conflict and war are generally perceived through three philosophic frameworks: realism, pacifism, and just war theory. Realism and pacifism are opposites of each other in many ways. To a pure realist, a state has no ethical constraints whatsoever: people, not states, are governed by ethics, and whatever a state does to advance its interests in times of existential conflict is appropriate. To a pure pacifist, no conflict or war can be ethical, so to discuss laws of war is to already be in unethical territory. A pure

[3] Although the Americans used conventional tactics far more often than usually realized.

pacifist might even oppose laws of war on the grounds that they are an attempt to make an inherently unethical activity more ethical, which is not logically possible. Furthermore, by making war less horrible, the laws of war may make it more likely or acceptable.

Very few people, and virtually no states, are either pure pacifists or realists. Most tend towards a third perspective, just war theory, which begins from the premise that war is part of the human condition. When it cannot be prevented, it should be conducted as humanely as possible, and combatants must take particular care to avoid unnecessary harm to civilians. Just war theory is tied to the state sovereignty model of international governance that reached its apotheosis in the post-World War II period marked by the founding of the United Nations in 1945. Membership in the U.N. and treaty and statutory obligations flow to states, not individuals, firms, or non-governmental organizations (NGOs).

The inviolability of state sovereignty, which underpins state-based theories about the conduct of war, has come under pressure from a number of developments. Among these are cases where states assert their right to intervene in another state where internal mistreatment of civilians is occurring; this has led to a new theory postulating a "duty to intervene," often called a "Responsibility to Protect" or R2P. Another example is the evolution of transnational networks of non-state actors, such as Islamic jihadists, which challenges the entire state-based structure. While these groups can claim sizeable membership, access modern weaponry, control territory, and even perform some of the duties of a traditional government—tax collecting, policing, education, sanitation, etc.—they are not states, and not controlled by states. While there have always been groups that use terrorism as a strategy, they were generally manageable as criminals or associated with a particular state sponsor. Global, informal, non-state terrorist networks capable (in theory) of obtaining and deploying weapons of mass destruction are a relatively new phenomenon, and one that challenges state-based legal regimes.

Extensive privatization of military activities presents another challenge. In today's confused conflict zones, like those of the recent wars in Iraq and Afghanistan, it is not unusual to find military personnel, private contractors, and espionage agents engaging in conflict simultaneously. Militaries may be bound by the laws of war and international treaties, but espionage agencies and private contractors are not (although they may be bound by their home country laws, or through contract clauses).

Finally, some technologies, such as the use of cyberweapons, raise a number of issues for the state-based model of international affairs, and more generally for just war theory. Because cyberattacks usually involve integrated networks that include components in many countries that are owned by many entities, the question of who can be attacked in response is a complicated one. Is it any company whose servers or network facilities were used, or any country that was transited by the attack? Cyberattacks are usually and intentionally difficult to attribute to any particular source with accuracy: does that mean that such attacks violate the Geneva Conventions' injunction against "perfidy" (deliberate deception regarding combatant status)? And when does a cyberattack justify a kinetic response? What makes a particular act—stealing large amounts of personal data from a foreign government, for example—a cybercrime versus a cyberattack?

These complexities do not mean that just war theory is obsolete, or an invalid source of guidance for applied ethical discussions of military and security technologies. But they do caution that we are a period of rapid evolution, both of the technologies and the laws and ethics governing them.

Just war theory can be broken down into three components, all of which combine norms, written laws and treaties, and customary international law.[4]

1. *Jus ad bellum* addresses the question of when a war can be ethically begun. *Jus ad bellum*, which is incorporated in the United Nations Charter, holds that in order for a war to be just it has to satisfy three conditions. First, it requires a compelling cause or justification, such as self-defense in the face of aggression. Second, a legitimate authority, such as a state, must make a public declaration of war. Third, the use of deadly force must be undertaken only as a last resort.[5]

2. *Jus in bello* addresses how wars may be ethically fought once they are begun. The initial Geneva Convention treaty of 1864 was the first to implement *jus in bello*. This began the development of what is often referred to as international humanitarian law, or IHL (IHL is often referred to as the "laws of armed conflict"). IHL has subsequently developed two major branches. The first deals with the need to separate combatants from non-combatants, and the protection of the latter to the extent possible. The second regulates the methods and technologies of armed conflict.

3. *Jus post bellum* is a much newer component of just war theory. It deals with questions arising from the termination of war, and subsequent peace agreements and associated activities (e.g., payment of reparations).

---

[4] Customary international law is a body of law that derives from custom rather than explicit treaty or statutory obligations. While not codified, it is often cited and relied on in national and international legal decisions and documents, although there is, not surprisingly, no universal agreement on specifics.

[5] While a pre-emptive attack is possible, there must be a powerful, immediate, and identifiable threat if such an attack is to be justified.

While these categories have traditionally been framed separately, they are obviously intertwined. An initially just war (say, responding to an attack in self-defense) may become an unjust war if fought using unjust means. Similarly, even if a war is begun unjustly, the parties are not thereby relieved from having to prosecute the war itself in a just fashion.

Theories like just war are an attempt to create a coherent framework out of the many principles, treaties, practices, and norms that have arisen over the centuries to regulate conflict. Because they arise frequently in the literature as well as in this book, the reader should be familiar with some of these general principles that inform just war theory.

Perhaps the most important basic requirement of the laws of war is *military necessity*: only those engagements and actions that are conducted in the pursuit of legitimate military objectives are ethical. This is obviously only the beginning of analysis, because many actions which might be militarily useful would still be unethical. Attacks on civilians, for example, might be argued by their proponents to be in pursuit of military objectives, but they would still be regarded under the prevailing interpretations of IHL as unethical and unlawful. The military necessity requirement is an important touchstone, in that it demands at least some effort to identify a military purpose for a particular action.

The core requirement of the principle of *discrimination* (sometimes referred to as *distinction*) is that combatants be distinguished from non-combatants, and that only valid military targets should be engaged. Applying this principle can be very difficult in counterinsurgency operations, where someone who is a civilian almost all of the time dons an explosive vest and becomes a terrorist.[6] This principle often arises in discussions of autonomous robotic systems. Some argue, for example, that robots will never be able to discriminate between combatants and non-combatants, and thus must necessarily violate IHL. Others, however, point out that in combat environments humans themselves are not terribly good at discrimination, and are prone to anger, fear, confusion, and a desire for vengeance that may lead to violations of the principle. They also point out that the assumption that robots can never develop such capabilities is only a hypothesis about a still-evolving technology, and thus may not adequately reflect future competencies.

The *proportionality* principle requires that the force exerted to achieve a legitimate military objective must not be excessive. A soldier cannot, for example, shoot and kill an individual who is yelling insults absent other factors, because the lethal response is far in excess of the amount required by the situation. Similarly, excessive collateral damage to civilian property, or excessive civilian injuries or deaths, is not permitted under this principle.

Many treaties and agreements such as the Hague Declaration prohibit "*weapons of a nature to cause superfluous injury or unnecessary suffering*." Treaties and customary international law, for example, prohibit use of such weapons as lances or spears with barbed heads, or serrated bayonets; expanding or explosive bullets; weapons with poison on or in them, and poison gas; biological weapons; weapons that produce fragments that can't be detected by X-ray; and blinding lasers. This principle clearly overlaps with the previous one: a weapon that causes unnecessary

---

[6] Whether terrorism is or should be subject to IHL in the first place, or whether it represents something other than war, is another complicated question.

injury and suffering usually violates proportionality as well.

Under just war theory and its current incarnation in the laws of armed conflict, collateral damage—harm to civilians and civilian property—is not unlawful per se. Under the *doctrine of double effect,* collateral damage is not unlawful if the causative combat action is otherwise permissible, and the military and not the collateral effect is the one intended. In a variant of the proportionality principle, the potential benefits of the combat activity must outweigh the collateral impacts. Thus, for example, it is not unlawful if a drone attack on a terrorist group kills a civilian, assuming that the other tests of lawfulness are met. Whether it is politically or strategically wise is a different question.

Finally, the principle of *command responsibility* makes officers responsible for the actions of those under their command. This principle, which will be very familiar to any officer in a modern military, ensures an identifiable chain of responsibility in case of any violation of the laws of armed conflict. It provides an incentive for those in charge to make sure that their personnel are trained in, and comply with, IHL.

While these principles sound simple, practice is often far more complicated. Command responsibility, for instance, is intended to protect lower-ranking soldiers from being unfairly blamed for poor training and orders that lead to inappropriate behaviors. But it doesn't always work that way. Some argued, for example, that the U.S. government violated this principle when low-ranking soldiers were disciplined—including with jail time—for the abuse of prisoners at Baghdad's Abu Ghraib prison. More senior personnel, including civilians in the chain of command, who encouraged or at least made such behavior possible, received only reprimands or demotions.

Further complicating the picture, these principles apply only to military personnel carrying out military missions. Today's combat environments can be much more complex, mixing elements of traditional military combat with policing activities and intelligence collection; private firms acting in many different capacities for many different employers; and individuals slipping from combatant to non-combatant roles unpredictably and without outward sign (e.g., donning explosive vests under civilian clothes). The opportunities for confusion are legion, and often cannot be reduced, in part because it may suit some of the combatants to encourage such confusion.

Concluding Observations

Today's mixture of emerging technologies, increasingly complicated combat environments, shifting geopolitical sands, and rejection of many of the international system's norms and values make the applied ethics and regulation of armed conflict a complex and charged domain. Nonetheless, without pretending to be exhaustive, I suggest here a few general observations.

1. **Law, applied ethics, and increasingly multicultural norms are co-evolving, interconnected, and in tension with each other.** Law is often established and relatively clear, at least in principle, and because it is a function of jurisdiction, it is conceptually apparent that most laws are not universal (even the U.N. Charter applies only to signatory states). Major treaties bind only ratifying states, and only indirectly or through domestic legislation bind individuals, firms, or NGOs. Norms are trickier, in that they are usually implicit and often assumed, especially by dominant powers, to be universal when they may not be.

2. **Hypotheticals can be powerful mechanisms to explore complex ethical questions raised by emerging**

military and security technologies, but they are not predictive. Military and security technologies are designed to cause damage and death, and are therefore quite scary. It is no surprise, therefore, that activists and others tend to generate dystopian hypotheticals about the development or use of these technologies (this has arguably been done by those seeking to ban autonomous weapon systems). This process can be helpful to policymakers and applied ethicists, as practice for thinking about scenarios and how to respond to them.[7] But the future path of any technology of sufficient power to be interesting is impossible to predict. How technologies interact with society is far too complex, and often too strange, to even guess at, and can only be adapted to in real time. The people that built the first computers had no idea how ubiquitous they would become, and the military researchers that built the initial Internet had no idea how we would be using it today. Public policy cannot be based on hypothetical scenarios because they imply a level of knowledge about the future that is impossible.

3. **"Coded language" is a way to avoid the analytical and rational discussion that is critical to applied ethics by directly engaging emotional responses.** For example, consider the coining of the word "frankenfood" by activists opposed to genetically modified crops, a brilliant public relations move. The integration of the Frankenstein narrative's implicit distrust of technology with something—food—that all people require is profoundly unsettling, and elides the underlying issues of science, technology, economics, and culture.[8] But the

---

[7] This is one reason why militaries, well aware of the chaos and unpredictability of conflict, conduct war games.

[8] Contrary to the implications of the term, studies have found no adverse human health or mortality effects from genetically modified food crops.

use of coded language does not, by itself, invalidate the underlying position: the applied ethicist has a responsibility to consider all facets of an issue as fairly, and with as much integrity, as possible, regardless of how inelegantly or even duplicitously they may be presented.

4. **Traditional disciplinary boundaries can be unhelpful when grappling with the ethical implications of emerging military and security technologies.** The initial decision to develop a technology will often reflect immediate uses and demands. A sophisticated analysis of the technology's implications must move far beyond that. But attempting to figure out a technology's impact through the lens of a single discipline—economic analyses, for instance, or engineering assessments—is to describe different parts of the elephant. The impacts of technology in society are so unpredictable and complex that an array of tools and disciplines are necessary to adequately assess them. Thus, for example, miniature flying robots might be quite useful in counterinsurgency environments, enabling targeted surveillance and attacks that dramatically reduce collateral damage. What do these flying minibots mean when they proliferate in civilian society, as they undoubtedly will, used by everyone from divorce lawyers to local news channels to political candidates?

5. **Embedded and unstated assumptions can predetermine the results of analyses.** For example, some analyses of unmanned aerial vehicle (UAV) technologies—drones—might assume that the drone-deploying power will always have air superiority (deployment of UAVs in Iraq and Afghanistan has been possible partially because the invading powers had absolute air superiority). This analysis would fail under easily foreseeable scenarios, such as a conflict between Europe and Russia, or the United States and China. All anal-

yses rely to some extent on assumptions; high quality ones will make them explicit.

Combat of various kinds is one of the few activities that have apparently characterized humans and their societies from the beginning of history. Identifying an objectively "right" set of values or principles to govern this most human of activities is impossible. A Chinese military planner, an ISIS commander, a harried U.S. platoon leader, an insurgent soldier, military contractor, an intelligence agent: all will have different perspectives on the applied ethics of their activities, some of which overlap, many of which do not. The voices engaged in these dialogs do not agree with each other, but in that they reflect the real world of conflict and emerging technologies—the real-world domain of the applied ethicist.

3

UNIVERSAL CONSCRIPTION AS TECHNOLOGY POLICY[1]

When the broad citizenry delegate the defense of their country to others, whether a small elite or a mercenary force, the nation often suffers. Rome fell in part because the Roman citizen no longer saw it as his duty to fight for his country, and after increasing internal instability and weakness, the external "barbarians" conquered. Machiavelli attributed part of the decline of 15th century Italian city-states to the rise of mercenary armies: city residents were willing to pay others to fight for them but not to assume the responsibility of becoming citizen soldiers. In the early 19th century, small professional armies, isolated from their citizenry, fell on the battlefield before the citizen armies of the new French Republic.

This historical trend today intersects with significant challenges. The industrialized nations are witnessing a military technological revolution of profound importance, one that increasingly shifts power from the human to the machine. At the same time, there is no shortage of high-profile strategic challenges: What is the appropriate strategy to stabilize the Middle East, or is that goal even feasible? Will the rise of China and concomitant shift in U.S.

[1] This chapter was co-written by Mark Hagerott, and an earlier version was published in *Issues in Science and Technology*.

geopolitical status be peaceful or violent? How will the global shifts in supply and demand for strategic natural resources influence geopolitics and future conflict?

For those concerned about the national security of the United States, these are important questions, but we believe that the attention they receive crowds out even more fundamental ones. How can the nation provide ethical and effective military, defense, and security capabilities in a period of unpredictable technological and social change? How can we train the technical workforce necessary to perform these functions and develop the institutional capabilities to shape and manage the weapons of a technological revolution that will rival the nuclear age in the depth and breadth of consequences?

The nation must address the problem of the rapid evolution of emerging military and security technologies, such as cyber- and robotic systems, especially given the larger context. In particular, the United States has completed the shift, initiated at the end of the Vietnam conflict, from a citizen to a professional military made up of a relatively small number of specialists and experts, who are now empowered to deploy and operate these powerful systems. This has led to a growing cultural gap between civilian society and the professionalized military, as fewer and fewer citizens and civilian leaders develop any familiarity with military culture and operations. This trend matters because the growing complexity of military and security technologies makes them less transparent to a democratic society and its institutions, even though society as a whole (at least in principle) has the ethical responsibility for the management and deployment of the tools of warfare.

In particular, we believe that the interplay between technological evolution and the shift to a professional military creates a deeply troubling dynamic. Most people already know they will not be exposed to the risks of conflict, given the professionalization of the armed forces. Now combine the absence of conscription with technological evolution that replaces humans on the battlefield with increasingly autonomous warfighting machines, and a fundamental social calculus—when to shift from diplomacy to war—may be altered in unprecedented ways. War should always be a last resort. Lowering the political and cultural barriers to initiating conflict is a dangerous development, all the more so if it emerges not from explicit and thought-out policy choice, but social and technological changes that no one is monitoring.

To be sure, hand-wringing about the trajectory of the military profession has occupied the energies of many a sociologist and historian, from Thucydides to Samuel Huntington; indeed, the relationship between civilian and military leadership is a theme dating back to Sun Tzu. These theorists have pointed to the risks of a military growing more technocratic and distant from "the people." But with the acceleration of technological advance and the dramatic increase in the complexity of modern conflict and the geopolitical context in which it occurs, a profound change is afoot—a change in both the nature of warfare and the relationship between American citizens and the nation's warfighting activities.

War Made Too Easy

Our thesis is simple. We believe it is neither socially nor technically advisable to rely on a progressively smaller group of specialists, increasingly separate from the rest of society, to provide the collective defense; nor does having such a small elite control the tools of modern, automated, and computerized war comply with democratic principles. Conversely, traditional military institutions and personnel are likely to be less than expert in conflicts where cyber infrastructure, financial networks, and other technological systems are targets for conflict. Under such circumstances, continued reliance on conventional voluntary military forces is unlikely to provide the expertise and skills necessary for comprehensive security.

We propose the return of universal conscription in order to provide for a more robust, sustainable, and democratic defense. Universal conscription is not intended to man the ramparts against a looming enemy wielding a numerically large field army, but to enfranchise the nation in the protection of its military capabilities, national security, and democracy, and to ensure appropriate and democratic responses to the challenges of increasing technological and geopolitical complexity.

We make two interrelated arguments in support of universal conscription. First, we need a broader cross-section of society in the military if we are to make better decisions about when, how, and at what scale to initiate conflict. Much of the thrust of modern military technology has the effect of reducing warfighter casualties. Unmanned aerial vehicles (UAVs) such as the Predator, for example, separate the airman physically from the battlefield, thus placing him or her at far less risk. Protecting military personnel from harm is necessary and desirable, but it may also lower the social, political, and psychological barriers to moving from negotiation to military engagement. We believe that broader social participation in the military could ensure an appropriate balance in democratic decision making about when to make the momentous transition to military action.

Second, technological competence is necessary, if not sufficient, for global power status, yet understanding and managing the complexity of accelerating technological change is challenging the ability of the United States, its citizenry, and its institutions. Military service, with its exposure to advanced technologies in relatively quotidian circumstances, can help create a more technologically sophisticated society.

At the same time, the accelerating evolution of technology across its entire frontier, driven by advances in nanotechnology, biotechnology, information and communications technology (ICT), robotics, and applied cognitive science, is challenging the adaptive capabilities of modern militaries. If the military is to be able to remain competitive globally in such a difficult and complex environment, conscription will be required to bring into the military a broader array of necessary skills.

For example, cyberconflict poses not just a technological and geopolitical challenge, but also a challenge to internal military culture. The geeks who, feasting on Coke and Skittles, are fearsome in ICT capability are not usually the kinds of personalities attracted to a traditional, strongly hierarchical, and heavily bureaucratic military organization. Nor do we expect that the institutional leaders, entrepreneurs, and change-makers who work with the geeks, and who understand the political sensitivities and social concerns about privacy, data management, open source, and the like, will be volunteering for military service. Appropriate management skills, in additional to technological capabilities, will be critical competencies for tomorrow's military. In discussing these issues, we will focus on the American context for the sake of simplicity and because we are more familiar with U.S. military and cultural issues. But we believe that these concerns will apply to many affluent nations fielding today's high-technology militaries.

Universal conscription—the compulsory enlistment of individuals, usually young men, in national service—has a long history. The Chinese Qin Empire, 221 to 206 B.C.E., used it; more recently, it was characteristic of post-Napoleonic European imperialism. In the early United States, militia service was frequently required, but conscription was generally not imposed.[2] Where conscription was imposed, as in the Civil War, it tended to be poorly administered and riddled with loopholes. With the advent of World War I, however, conscription was modernized by the Selective Service Act of 1917, and it was implemented again in World War II. Unlike the Civil War era, conscription for these wars was not politically contentious.

This changed with the Vietnam War. In fact, one of the most significant contributions to a reduction of domestic social tension in the United States as the Vietnam conflict wore down was the ending of the draft in 1973.[3] Although some countries, such as Israel, Iran, Switzerland, Turkey, and South Korea, still rely on universal conscription, the U.S. shift to a so-called "all-volunteer army" is the dominant trend, especially in Europe. France, Germany, Italy, Sweden, and the United Kingdom, for example, have all suspended or eliminated conscription in favor of volunteer militaries.

And indeed there is an obvious political benefit in volunteer forces, where those who serve, and often their families as well, have knowingly accepted the risk of going to war. In contrast, universal conscription means that everyone of a certain age could potentially serve in the military, and warfighters are drawn from a broad cross-section of society. This spreads the risks associated with military activity and thus makes decisions about the deployment of forces much more contentious, which is not necessarily a bad thing.

---

[2] Article 1, Section 8 of the Constitution separates the power to support the Militia, administered at the state level, from the power to raise and support the national Army and Navy through, for example, conscription.

[3] The Selective Service system, which is the administrative backbone of the draft, remains active.

In addition, it is far easier to manage an all-volunteer force. Draftees in a country such as the United States are opinionated, stubborn, cantankerous, questioning, and difficult to lead, as one of us (Allenby) learned first-hand as an Army officer during the Vietnam War. As Robert Goldich, a noted expert on conscription has observed, "draftees did not internalize the norms and psychology of the career force; but rather accepted them, externally and reluctantly, and adapted as best they could. Today, broadly uniform attitudes permeate the entire force, from private to full general." Is this monolithic culture, unchallenged by the draftee, a good thing in the long run?

Some argue that, in a melting-pot culture such as that of the United States, class and regional differences become far deeper and more difficult to manage when universal service is eliminated, since it is one of the few arenas in which citizens are mixed without regard to their differences. Moreover, civilian leaders who have not served are necessarily not as intimately familiar with the culture, operations, strengths, and weaknesses of military institutions. As journalist Dana Milbank noted in a 2012 *Washington Post* op-ed, only 86 members of the House and 17 senators have served in the military, a rate of only 19% – the lowest since World War II (the high, in contrast, was 77% in 1977–1978). Are our legislators more likely to make mistakes in the deployment of U.S. forces? The Iraq War is often used as an example of such a mistake.

Without anyone intending it, the elimination of universal service deepens the gap between military and civilian culture, a gap that some believe has grown to a chasm. In an October 2011 speech at West Point, former Secretary of Defense Robert Gates observed that "there is a risk over time of developing a cadre of military leaders that politically, culturally, and geographically have less and less in common with the majority of the people they have sworn to defend." And, in fact, the Reserve Officers' Training

Corps (ROTC) programs, which are a major source of military officers, are disproportionately concentrated in the South or Rocky Mountain states, whereas the populous East and West coasts are underrepresented. As Gates noted, "Alabama, with a population of less than 5 million, has 10 Army ROTC host programs. The Los Angeles metro area, population over 12 million, has four."[4] Gates further observed that "it is off-putting to hear, albeit anecdotally, comments that suggest that the military is to some degree separate and even superior from the society, the country, it is sworn to protect."

At the same time that mass societal participation in the military has abated, military technologies have evolved to reach out, identify, observe, coerce, and in some cases kill, human targets. These technologies have empowered a progressively smaller numerical group, such as a team sitting in a control room in South Dakota directing an unmanned Predator. Alienation is a hazard, both from the human enemy and fellow citizens, who have fewer and fewer connections with these isolated technocratic warriors.

Thus we see the isolation of the warfighter driven from two directions. The first involves, as then-Secretary Gates explained, "a series of demographic, cultural, and institutional shifts that have made the military less representative of the American population as a whole, mostly as a consequence of ending the draft." The second direction is from the momentum of technological change, perhaps trending toward a future battlefield, described in Antoine Bousquet's 2009 book, *The Scientific Way of Warfare* (required reading at the Army Command and Staff College),

---

[4] The National Service academies at Annapolis, West Point, and Colorado Springs are exceptions in this regard, drawing by law from every congressional district precisely with the intent of providing representation aimed at a politically plural officer corps.

in which "swarms composed of millions of sensors, emitters, microbots, and micro-missiles and deployed via prepositioning, burial, air drops, artillery rounds, or missiles, saturate the terrain of conflict."

The New Ways of War

If the political forces that ended conscription are fairly clear, what about the forces of sociotechnical change leading the military toward the development and adoption of new military and security technologies? The trend—and its consequences—toward drones, cyberweapons, and autonomous vehicles and robots that put military personnel increasingly far from the action are apparent just from reading the headlines about current conflicts. What is driving this change?

The U.S. military is expected by the public and civilian leaders to project American power around the globe, but with minimum casualties. This is a radically new perspective on military operations. Throughout history, and reaching a peak in the World Wars of the 20th century, such operations often killed tens of thousands in a single battle. But what allowed this new perspective to become plausible? First, of course, technological advance, which now creates the political expectation that wars can be fought with few or even no casualties. But there is another reason: budgetary pressures, an aging population, and changing social norms mean that the military must plan to be smaller, meetings its needs with fewer young recruits, and that competition with private firms for those individuals will become much more intense. These stresses demand increased labor efficiency, or more mission accomplishment per unit of warrior. Substituting robots for warriors doesn't just save lives, it substitutes capital for labor and enables performance even if labor pools shrink.

Taken together, these conditions strongly suggest that the challenges posed by accelerating technological evolution to military and security institutions, and to society generally, will only continue to increase. Although some say that such a pattern of innovation has existed for thousands of years (say, back to the Trojan Horse), their assumption that nothing fundamental is different this time is highly contentious. The continuing evolution of computational capabilities, the creation of increasingly intelligent and mobile machines and networks, and now initiatives to radically redesign the warfighter himself or herself, are each individually significant, but when taken together create a discontinuous complexity.

The huge flows of combat-critical data, the increasing technological autonomy of modern weapon platforms, the changing relationship of the soldier to the battlefield as U.S.-based warriors direct drones around the world—all of these come together and are rapidly coevolving, in ways that raise challenges that are new and not yet well managed. The need for both military and civil society institutions to emphasize technological knowledge and sophistication will need to accelerate greatly if they are to remain viable and competitive. But such pressures also suggest that the gap between military and civilian understanding of, and comfort with, technology is liable to grow, as will the cultural and communication gap between those sectors.

Why is this gap a problem and how might we manage or reduce it? Technologies such as UAVs being flown over Afghanistan today by pilots stationed outside Las Vegas certainly reduce American casualties. Replacing special operations warriors with robots, as opposed to augmenting their capabilities in the field, is a long way off, but it is not difficult to see that reducing the potential for U.S. casualties makes military action less politically risky and thus lowers the high barrier against going to war.

Universal Conscription

Going to war should always be a last resort for a civilized society, and never undertaken lightly. Iraq is a cautionary tale, but it pales next to the reduced disincentives for conflict that foreseeable technological evolution, especially in the continued absence of conscription, may soon create.

If there is a high probability that foreseeable technological evolution will make war more socially acceptable, we obviously need other ways of making war less desirable. Neither the speed nor direction of technological change is likely to raise new political barriers to warfare. Thus one must look for mechanisms for managing and maintaining barriers to conflict in other domains, such as public policy. Our argument is that universal conscription—a properly designed draft that exposes all individuals of a certain age to the potential of becoming involved in the military during a conflict—can have the effect of building back into society an appropriate conservatism about engaging in warfare that technological change has in part undermined.

This effect operates even when technologies may be removing many soldiers from traditional conflict zones. The global nature of terrorist networks and the extension of weapon systems and warfighter participation across regional and global scales mean that violence might well follow the soldier to wherever she or he is stationed. There is evidence as well that UAV pilots, despite being physically removed from combat, still can suffer from post-traumatic stress disease. A draft would not only spread the risk of combat injury or death across a broader, more diverse population, but it would also expose many more people to the possibility of being drafted, thus giving them and their families a direct interest in national decisions about when to go to war. A well-designed conscription program can, therefore, reduce incentives for violence to resolve geopolitical differences.

Simply put, if technology is making war too easy, the draft is one of the few ways to keep it hard.

A Few Good Nerds

If the most immediate and direct benefit of a draft is to counterbalance the possibility that technological change makes military action too politically easy to pursue, then a secondary benefit is to enhance the capacity of the military, and even U.S. society more broadly, to understand and manage rapidly evolving technology systems. The unpredictability and complexity of technological change challenge military organizations in several ways.

One is the challenge of attracting and keeping necessary technological talent. This can be done piecemeal. For example, Admiral Hyman G. Rickover, the father of the nuclear navy and the commercial nuclear reactor, when he faced the accelerating technological changes of the early nuclear age, lamented the shortage of technically qualified sailors and officers entering the service. To tap the talent in civil society, he went to extraordinary lengths to create innovative programs that permitted qualified civilians to enter directly into his organization, thus attracting recruits who would have normally shunned military service.

Today, the challenges arise not just in a specific area that one strong-willed individual can address, but across the technological frontier. A policy of relying on ad hoc and idiosyncratic fixes, although perhaps necessary in the short run, is doomed to inadequacy. Contractors may be a short-term option, but their incentives are economic and not institutional and patriotic, so their reliability, especially under combat conditions, may not be the same as with warriors. The necessary systemic policy fix could be universal conscription. Designing a conscription system that is able to mitigate the issues we've identified, and that is also perceived as fair and politically acceptable, is admittedly no small task. But the costs of not doing so are high and increasing.

A related challenge is simply keeping track of emerging technologies and understanding their potential implications for military operations and national security. The military is well aware of this challenge. At the U.S. Naval Academy in Annapolis, for example, every midshipman, regardless of academic major, is required to take core courses in cyberstudies and enough electrical and computer engineering that he or she can understand the emergence of these complex systems. Such experiences demystify technology and enable students, and those they lead, to be more realistic, and more grounded, in an environment of rapid and unpredictable technological evolution.

Relying on a small, volunteer elite—as the military does now—to manage major technological revolutions across virtually all security domains is unrealistic. A self-selected volunteer elite, no matter how competent, will not reflect the skills and, more importantly, the perspectives, cultural competencies, and implicit knowledge embodied across the society. Will enough geeks volunteer? Will enough experts in finance, who can help protect critical assets from unrestricted cyberwarfare, be available and aligned with more-traditional military defense institutions? National security by definition reflects the values and interests of society as a whole, but even as the complexity and capability of military and security technology accelerates, and difficult policy debates about UAVs, surveillance, and global wars on terrorism flare up, military and security competencies are devolving to a relatively small and increasingly less representative volunteer elite.

This is not only likely to leave the military technically deficient, as inadequate numbers of those with important skills (hacking competencies, for example) fail to volunteer, but it also creates a challenge to democracy. Conscript armies are a cultural, socioeconomic, and geographic cross-section of society. After a year or two, conscript soldiers or sailors go home, bringing with them experience with the complex and evolving roles of technology in the nation's security enterprise. They raise families in the communities from which they came, communities from across the society that are now enfranchised in decisions about the military, because it is their children and friends that are involved.

Faced with the tsunami of innovation across numerous fields unleashed at the beginning of the 21st century, technologies are shaping our military enterprise in ways that may radically challenge our social capacity to govern that enterprise in a manner that is technically competent and democratically responsive. A self-selected volunteer elite, no matter how competent, will not reflect the skills and, more importantly, the perspectives, cultural competencies, and implicit knowledge embodied across the society.

A properly designed and broadly implemented universal conscription program offers a policy approach that can address both elements of this challenge: expanding the technical proficiency necessary for tomorrow's national security, and enfranchising a more diverse and inclusive cross-section of society in the process of national security provision and decision making. No doubt piecemeal approaches may be offered to address the influences of new technology in the military and security spheres. But the current status quo of a volunteer military fails to provide our military apparatus, and our democracy, the capabilities needed to successfully navigate the world we are busy creating. History teaches us that we should not insulate our military from the citizenry, and we should not let the marvels of technology blind us to this important lesson.

4

CYBERCONFLICT AS STRATEGIC FRONTIER

"In my opinion, it's the *greatest* transfer of wealth in history."

—General Keith Alexander, National Security Agency director and CYBERCOM commander, on the billions of dollars of private intellectual property stolen by suspected Chinese hackers.[1]

Introduction

Students of strategy will be familiar with the two individuals who did the most to create the field of naval strategy: Alfred Thayer Mahan and Julian S. Corbett. Mahan, in books such as *The Influence of Sea Power upon History, 1660-1783, The Influence of Sea Power upon the French Revolution and Empire, 1793-1812,* and *The Interest of America in Sea Power, Present and Future,* created an influential strategic framework for war on the oceans. Corbett, in works such as *Maritime Operations in the Russo-Japanese War,*

[1] As quoted by Tabassum Zakaria and David Alexander, "U.S. spy agencies won't read Americans' email for cybersecurity," *Reuters* (9 Jul. 2012), available online: http://www.reuters.com/article/net-us-usa-security-cyber-idUSBRE86901620120710 (emphasis added).

1904-1905, and *Some Principles of Maritime Strategy,* created an overarching strategic framework for how naval warfare and operations integrated into overall military and strategic operations. In essence, they took centuries of commentary regarding military operations on land (ranging from Sun Tzu to Carl von Clausewitz and Antoine-Henri Jomini), modified it, and applied it to a new domain: the ocean.

Cyberspace is the new domain of security and military conflict, and it raises similar challenges. We do not lack authoritative and very valuable studies of cyber conflict, such as the 2013 NATO *Tallinn Manual on International Law Applicable to Cyber Warfare*. Nor do security experts fail to understand the importance of cyber, especially as applied in a strategic environment emphasizing New Generation and Unrestricted Warfare.

Nonetheless, a clear, robust, and workable framework for developing and understanding cyber conflict in all its aspects, analogous to that developed by Mahan for naval operations, has yet to emerge. Similarly, while the use of cyberforce either as a coercive mechanism (e.g., the alleged attack by North Korea on Sony Pictures in 2014) or as a component of major projections of force (e.g., the 2014 Russian attacks on Georgia and Crimea, or the U.S. attack on Iraq in 2003), is well established and, indeed, almost traditional at this point, an overarching strategic framework analogous to Corbett's has yet to be developed for the cyber domain.

Cyber is a challenge that cannot be addressed simply by extending previous military strategic thinking, as the naval theorists were able to extend land warfare strategy to the oceanic realm. The effort to treat cyberspace as simply another military domain by extending slightly modified existing strategies into the cyber realm is proving difficult. That does not imply that all previous experience and strategic theory is inapplicable; the challenge is to balance learning from the past with understanding where new strategic thinking is required. The strategist must learn from the present, including events such as the Russian invasion of Ukraine—where Russia practiced a somewhat experimental version of information warfare that included significant cyber components—and from future scenarios, which are especially valuable when technology is changing rapidly and unpredictably.

Cyberspace is a realm that, while useful for traditional combat operations, is suited for conflict at all levels, from criminal activities to broad cultural competition, conducted by adversarial states, non-governmental organizations, insurgencies, cultural movements such as jihadist Islam, quasi-governmental private entities, and others. Cyberattacks avoid the direct and explicit impact of kinetic aggression, enabling long campaigns of subterfuge and stealth sabotage. They can be used to degrade all elements of a competitor's society, from the performance of financial and transportation networks, to subtle manipulation of social networks and media feeds. Cyberattacks can be the work of criminals or state-sponsored groups, with each hacking incident chipping away at trust in critical information systems while potentially enriching bad actors. Cyberattacks are especially effective in such cases because sophisticated attackers are anonymous.

As these various activities imply, cyberattacks occur in multiple domains and for a number of reasons, but always involve the use of information systems to execute (or to embed the ability to execute at some future time) a destructive action. The underlying motive may involve military, espionage, political, economic, or criminal goals, and will often be mixed.[2] A competitor may resort to cyber

---

[2] Russia, in particular, has tolerated and indeed encouraged cybercriminal enterprises for strategic purposes. In part this is because Russian doctrine draws on a long history of revolutionary

methods due their relatively low human, economic, and political costs, and high returns. That is, cyberattacks have the potential to return a great deal of information and create broad disruption—particularly in the realm of military command, control, communication, and intelligence (C3I)—for a small initial investment, all while providing the perpetrator with a veil of anonymity. Cyber conflict is thus the ideal form of asymmetric warfare, used by the relatively poor and powerless against the wealthy and dominant.

Technically, cyberattacks of any type may take advantage of a number of technical weaknesses. These include errors in software and network design not previously identified (zero-day attacks) and system vulnerabilities (e.g., distributed denial of service, or DDoS, attacks and hacking). They may take the form of deceptive practices, as in the case of phishing scams, malicious links, or Trojan horse programs. They might seek to immediately disrupt a target system, or to embed "logic bombs" or other devices that can be triggered in the future. With respect to security breaches, humans are often the weakest links, since they share USB drives which can harbor viruses, give up passwords, and carelessly install malware.

Law enforcement is often an inadequate response to cyberattacks. Unless mandated by law and regulation, many companies fail to report cyberattacks and significant data losses because of the stigma they fear will attach to them if they do so. Thus, law enforcement officials might not even find out about many successful attacks. Even when they do, domestic law enforcement in many cases will not have appropriate tools to investigate, and bring legal proceedings against, cybercriminals who frequently operate from different jurisdictions (assuming they can be

---

action against established authority, leading to a distinctly utilitarian approach towards cybercrime.

identified in the first place). It is unlikely that a Russian cybercriminal who successfully raids an American financial institution would be subject to any sort of legal repercussions.

While private industry has assumed the responsibility to protect their systems, in many cases the financial losses due to information security lapses often amount to less than the cost of security to prevent such losses. This means that there might be a "public good" dimension associated with cybersecurity in that the investment to protect electronic assets, though it would benefit everyone, is economically under-incentivized for individuals and firms. Cyberspace, as a result, remains the digital equivalent of the Wild West.

Note that the terms that are used in this domain—cybercrime, cyberattack, cyberwar—tend to reflect the subjective intent of the aggressor and the mechanism by which society responds,[3] rather than the nature of the technology used. This is not the case with other domains: it is very hard to see an incursion by a tank battalion as anything but a military event.

The domain also enables a level of strategic mischief that other technologies cannot support. If I want to undermine an economy, and I can do it through a series of individual attacks, each one of which is treated merely as a criminal event by the victim, I can have strategic impacts without my competitor even understanding that I have engaged him at a strategic level. The basic idea is not new: the Soviets and Americans played such games in the Cold War. What is new is the potential scale and scope of such activities enabled by cyber technologies.

---

[3] For example, is a particular transgression addressed by the criminal law system, or the military and security forces?

Cyber is a domain in itself and a source of weapons and capabilities that are frequently deployed as part of an integrated warfare strategy, as Russia did in Crimea, or as a complement to conventional warfare, as the United States did in its Iraqi wars, and Russia in its war with Georgia. ISIS has used social networking and Internet capabilities effectively in both establishing its brand and recruiting new members. Some discussions of cyber include questions of artificial intelligence, both as part of cyber systems (e.g., as a mechanism internal to networks that helps detect and respond to cyberattack), and as enabled by cyber technology (e.g., AI analysis of big data to identify terrorist threats).

This points to another difference between cyber and naval conflict. Naval activities have an intuitive boundary (military operations in or near navigable waters), whereas cyber appears in many different ways, in many different guises. One of the major complications in nascent efforts to regulate cyberwarfare is precisely the challenge of identifying boundaries around what activities and technologies are going to be included in the regulatory regime.

Cyber, and the Laws and Norms of Armed Conflict

Whether existing norms and frameworks adequately manage a technology is a pretty good test of how fundamentally new a technology really is. From this perspective, a number of factors indicate that cyber is indeed a radical new technology.

The following table, based on NATO analyses that are only a few years old, suggests a number of norms that, even in a short period, appear either overly optimistic or outdated. For example, "responsibility" is not nearly as simple as suggested: what if the attack has been sponsored through a private entity precisely to enable denial of responsibility? And the idea that a state or non-state entity would provide early warning of a cyberattack clearly doesn't apply if the state is using that attack as part of its strategy. More subtly, different countries disagree on what kinds of information constitute attacks. In the United States, for example, stealing commercial data, especially from defense firms, would be the kind of attack that politicians and military strategists would worry about. But in countries such as China and Russia, tools that circumvented government controls on information might be a more egregious attack on the state. Different cultures, in other words, have very different ideas about norms governing information generally, and any effort to regulate or manage the cyber domain will fail unless it recognizes, and can somehow manage, those fundamental differences.

The more significant question is how cyber as a technology might affect the laws of war. As noted in Chapter 2, the laws of war are now and have been very important ways in which the horror of combat has been managed, especially as regards civilians. To the question at hand,

> *If global strategic confrontation is shifting to a multidimensional conflict model, and cyber warfare is the emblematic technology for this kind of conflict, there is necessarily a disconnect between a legal structure designed for traditional military operations and the novel implications of cyber technology.*

Cyber technology undermines some of the most important questions an actor considering military action must answer. By tradition and by law, a just war requires attribution: you need to show that you were attacked, and that the attack came from the entity against which you propose to initiate military operations. It requires public declaration: you need to be clear that you are, in fact, going to war (a requirement that alone renders Russia's attack on Ukraine illegal). It must be proportional: you need to show that going to war is a reasonable and balanced response.[4] All of these requirements are challenging to apply to cyberattacks that are, by design, difficult to attribute, designed not to justify the public declaration of war, and often don't cause the kind of damage for which a declaration of war is a proportional response.

Table 1: Evolving Cyber-Norms

Norm	Description
Territoriality	Information infrastructure located within a state's territory is subject to its control.
Responsibility	A cyberattack launched from an information system located in a state's territory invokes the responsibility for the attack.
Cooperation	The fact that a cyberattack has been conducted via the information system located in a state's territory creates a duty to cooperate with the victim state.
Self-Defense	The right to self-defense when facing a clear and imminent attack is guaranteed.
Data Exchange	Information infrastructure monitoring data is perceived personal unless provided for otherwise.
Duty of Care	The responsibility exists to implement a reasonable level of security in relevant information infrastructures.
Early Warning	The potential victims of an incoming cyberattack should be notified.
Access to Information	The public has the right to be informed about threats to their lives, security, and well-being.
Criminality	Every nation is responsible for including most common cyber offenses in its substantive criminal law.
Mandate	An organization's capacity to act (and regulate) derives from its mandate.

Based on Enekin Tikk, "Ten Rules for Cyber Security," *Survival: Global Politics and Strategy*, vol. 53, no. 3 (2011): 119-132.

[4] Japan attacking Pearl Harbor is clearly a justification for declaration of war; a criminal attack on a private firm's database of charge card transactions is not, under current law.

There are, of course, conditions under which a cyberattack may rise to the level of an "act of war," justifying initiation of military operations in response. The best recent example is the Stuxnet attack on Iran, the release of a computer virus that targeted centrifuges used to enrich uranium to the high level necessary for use in nuclear weapons. The apparent goal was to shorten centrifuge lifespans and, ultimately, delay Iran's nuclear weapons program. While Iran accused Israel and the United States of developing Stuxnet, no proof was ever presented—although both countries implied that they were indeed the source of the virus, since their strategic interests lay in presenting themselves as technologically superior to their weaker Iranian foe. Despite undeniable physical damage to critical Iranian assets, no party declared war.[5]

This example illustrates the uncertainty as to what constitutes an "act of war" (*casus belli*) in cyberspace. Stuxnet was active, in that it caused physical damage as part of its operation; suppose the attacking parties had instead planted a logic bomb (software with the capability to cause damage if triggered, but not causing any physical damage in its inert state in the system) in the centrifuge software, and threatened to detonate it? In that event, there would be no damage and, at least in the physical world, no harm. Could malware ever constitute "arms"? At what point does a cyberattack become an act of war?

Another example of cyber conflict occurred on September 6, 2007, when the Israeli Air Force bombed a nuclear site under construction at Dayr al-Zawr in Syria. Apparently having purchased access to software "backdoors" in a radar system that the Syrians had purchased from France, Israel allegedly disabled the system that

---

[5] At least some experts believe that Iran could have done so, had it wanted to, and had it been able to present solid evidence that the United States and Israel were the attacking parties.

would have alerted Syria to the incoming Israeli planes.[6] The Israeli force slipped past Syrian air defenses, bombed the target, and exited without further incident.

At what point in the Dayr al-Zawr incident did an actionable act of war actually occur?

- When the backdoor was coded;
- When the compromised computer component was installed;
- When the backdoor was sold to Israel;
- When the malware was introduced into the Syrian air defense system;
- When the malware was actually activated;
- When the physical system was damaged; or
- Never.

It also raises the question of whom, exactly, Syria should respond against, assuming the current hypothesis is correct: Israel, who exploited the software backdoor, or France, who allowed the backdoor to be created in the first place?

These are, of course, not just technical legal questions, but also practical. Even if they had the legal authority to declare war on the alleged aggressor, in both the Stuxnet and Syrian radar cases, neither victim did, even though attribution was reasonably clear in both cases. Retaliation is a political decision, and like all such decisions, it is heavily dependent on context. Analogies to current practice can help us think through some of these questions. For example, is setting up logic bombs in an adversary's critical cyber-infrastructure conceptually the same as establishing a network of missiles that are targeted to physically attack that adversary? If it is, then simply implanting logic bombs would not be an act of war, but activating them might be, depending on the effect.

---

[6] Again, there is no public proof enabling attribution.

It may be broadly true that recovery from a software attack is usually easier and quicker than recovery from the physical destruction of an asset, but this is not always the case. Certainly in both of these instances, at the point that physical damage occurred there was little practical difference between the software attacks and using kinetic means to disable the radar or damage the centrifuge system.

But suppose that Stuxnet had brought the centrifuge system to a complete halt, without doing any physical damage to the system. It is unlikely that the response in that case could have been initiation of war, because under the United Nations Charter, parties are prohibited from using force unless they are either responding to an "armed attack" or their use of force is sanctioned by a Security Council resolution, neither of which would obtain. Any response, whether conventional or cyber, has to be proportionate to the injury, and employ only as much force as is required to attain a legitimate military objective, which makes a legal military response difficult in any event. Finally, under applicable international humanitarian law, responses cannot deliberately target civilian noncombatants or civilian "objects," although of course this is what cyberattacks often do.

In evaluating the implications of cyberconflict, it is worth emphasizing that the norms and legal requirements of the laws of war apply to military actions, not to espionage. This becomes pertinent because the cyber domain can be extremely valuable for espionage operations; the ambiguity of the context translates into ambiguity of the applicable legal regime. For example, cybersecurity firms believe that the Flame virus, a malware program discovered in 2012, succeeded in effectively turning personal computers throughout the Middle East into listening devices. The program could record keystrokes, audio, screen shots, and network traffic on an infected computer, and send that information to a central server. Yet this particular malware has not generally been considered by cybersecurity experts to constitute a cyberweapon in the military sense, since its actions constitute espionage, not an armed attack.

Another legal issue arises from the laws against so-called "treachery and perfidy," terms referring to acts in which an enemy combatant is tricked into trusting an agent with the intent to betray that trust. Examples including waving a white flag of surrender while continuing to fight or pretended to be wounded in order to gain a military advantage. Cyberconflict is full of such tricks—indeed, the core of the concern about attribution is precisely that it enables such trickery.

Some analysts have suggested that all uses of cyber weapons constituted perfidy, inasmuch as their origin, exact target, and purpose are clothed in subterfuge. This is essentially an argument that the existing laws of war are obsolete when applied to cyberspace because of the characteristics of the technology itself. Cyber warfare is arguably perfidious in two ways. First, it is perfidious in itself as an unattributed, and hence deliberately misleading, weapon. Second, it is an enabler of military campaigns and activities that are, in themselves, perfidious as well, such as the Russian invasion of Ukraine.

The history of the laws of war embeds another basic norm: protecting non-combatants to the extent possible. Cyberconflict challenges this norm in two ways. First, cyber warfare easily and effectively targets civilian institutions, networks, and operations. A rifle shot on a battlefield has an effect in that environment alone; a virus released in a military computer system or as part of an attack on a country will expand without regard for the boundaries between civilian or military systems.

Second, the global connectivity of the Internet and the nature of cyber operations make distinguishing combatants from civilians challenging. Could a computer hacker developing a program that can be used as a weapon be considered an enemy combatant? What about the Internet service provider that hosts the software? What about the owner of the communications infrastructure over which the software is distributed?

These questions become more difficult if cyberspace is understood to be an arena for cultural competition and conflict. For example, the People's Liberation Army strategic analysis *Unrestricted Warfare,* discussed in Chapter 1, identifies George Soros as a terrorist.[7] Soros, other financiers, and financial institutions all operate across vast networks that integrate across many companies, countries, and sectors. Are the firms and individuals in such networks to be considered enemy combatants under an Unrestricted Warfare strategy? Is a cyberattack on this constellation of individuals and organizations a legal attack on combatants, or is it unacceptable targeting of civilians?[8]

A final set of questions arises around cyberconflict and the just termination of conflicts. For example, does, or should, international law require the use of "ethical software," equipped with some sort of "off switch" that can facilitate termination of conflicts in a just and permanent way? Should cyberweapons be required merely to encrypt data rather than deleting it, so that the key can be turned over to the enemy upon resolution of the conflict, thereby

---

[7] See Qiao Liang and Wang Xiangsui, *Unrestricted Warfare,* U. S. Central Intelligence Agency translation (Beijing, China: People's Liberation Army Literature and Arts Publishing House, 1999).
[8] Terrorism and counterterrorism taken together form another domain where such questions are legion.

keeping permanent collateral damage to a minimum?[9] The complexity of information systems might make policing such agreements impracticable.[10] For the smart competitor, there are also strategic opportunities in these questions: one might, for instance, encourage an opponent to halt operations by building reversible software weapons, and threatening to make them irreversible if rapid agreement on terms isn't achieved.

A Last Scenario: Cyber, Civilizational Conflict, and Death by 1,000 Cuts

The development of strategies that dramatically expand the scope and scale of warfare leads to the possibility of a "death by 1,000 cuts" scenario. The idea is that a patient and well-organized attacker wishing to avoid conventional or nuclear confrontation with an adversary would, instead, initiate a campaign of inflicting small and incremental damage on financial, technological, social, institutional, or other systems by hackers over a long period of time. This could be a very effective mechanism for asymmetric warfare, weakening a more powerful opponent over perhaps decades, and eventually leading to the relegation of the adversary to second-class status without a need for direct military confrontation.

Such a campaign is a natural fit for cyberspace, where criminal activity is prevalent and substantial impacts can arise from a series of small operations over a long period of time (as General Alexander notes with reference to Chinese cyberattacks at the beginning of this chapter). Choosing the right targets, such as financial institutions or valuable intellectual property, can enable an adversary to

---

[9] There are software systems that do such a thing now, which are usually used for criminal purposes.
[10] How can one verify that all logic bombs have indeed been dismantled, especially when they needn't be local?

impose substantial costs without great military risk; in fact, the U.N. Charter explicitly states that economic actions do not warrant an armed response. This is arguably another case where technological advance has undermined the assumptions underlying the law. Reinterpretation of it may allow for responsive actions such as the imposition of sanctions on one state by another for cybercrimes, but the Charter clearly does not contemplate a long-term cyber campaign by parties unknown against the information infrastructure of a modern state.

Conclusion

The value of information for military operations has been known since antiquity.[11] The manipulation of information for strategic ends is not new either. What may be novel now is the sheer volume and velocity of information flow, and the accelerating integration of cyber networks—and thus their weaknesses—into designed and built systems at all scales. The challenge with cyberconflict, then, is not simply to invent an entirely new strategic framework as if past experience is no longer applicable. Rather, it is to determine what elements fit into existing categories, norms, and laws; what elements are different enough to require new agreements, strategies, and doctrines; and what elements are so unprecedented that they destabilize or render obsolete existing norms, laws, or institutions. Cyberconflict necessitates the establishment of a governance regime in a world where unrestricted warfare, and competition and conflict at the cultural level, are inherent in the strategies and doctrines of Russia, China, and other powers, including non-state entities such as ISIS.

---

[11] Sun Tzu's *The Art of War*, for example, devotes an entire chapter to spies, spying, and the importance of good intelligence.

5

ETHICAL CHALLENGES OF MILITARY ROBOTS

Introduction

The word "robot" was first introduced in a 1921 science fiction play, *R.U.R*, by the Czech writer Karel Čapek. The concept and the term immediately became popular, and by now most people have an intuitive understanding of what a "robot" is. But like many intuitive terms, that doesn't mean we really have clear understanding of it. Humanoid robots developed by Honda and others represent the conventional view of a robot as a mechanical person. But a broader way to think about robots is as artifacts that have the ability to sense the environment, and to effect change in that environment based on that sensory input. Hence the 1939 New York World's Fair, with themes that included robots and labor-saving devices, displayed a toaster as, in fact, a robot.

By the same token, modern land or sea mines are arguably not only autonomous robotic systems, but *lethal* autonomous robots. A mine monitors its environment, responds to input based on that monitoring by remaining dormant or exploding, and generally does so without any immediate human guidance once it is deployed. Mines are robots, just not mobile ones; they operate without direct human direction (as compared to, say, a drone attack), and they are capable of—indeed designed for—lethality.

Some argue that a robot must be a machine; others are not as restrictive. Is a software device that is entirely virtual (a "bot") a robot or not? When individual devices are deployed in a battle zone, but connected together into grid computational systems from which emergent behaviors can be generated, which is the robot: the individual device, or the network? When one integrates brain tissue into an otherwise mechanical system, is the result a robot, or is it something else? Many militaries are experimenting with integrated techno-human systems to evaluate large and complex information inputs from battlefield sensors in near real time, with the output emerging from the system as a whole, not any particular individual. Is such a system to be thought of as "human" or as "robotic"—or neither?

The definitional issue may sound arcane, but it is in fact central to debate about how to govern robots. Today, there are some calls, especially in Europe, for precautionary banning of lethal autonomous robots. But if there is no accepted definition for such a category, it is unclear exactly what is at issue. Drawing up any sort of legal document becomes very difficult, since no one knows exactly what is being regulated. Questions of definition thus must be resolved as experts think about how robots of various kinds fit into the existing framework of the laws of armed conflict, or whether new considerations are necessary.

Modern military and security robotic systems operate either under direct human control or in functional spaces that are tightly bounded. The Predator and the Reaper drones deployed by the United States are examples of relatively traditional, human-controlled robots, as are the explosive demolition robots deployed to Iraq and Afghanistan. Military and civilian drones, from the small quadrotor helicopters available from Amazon.com used primarily by hobbyists to the large unmanned platforms used by the military, are an increasingly familiar type of robot.[1]

An example of a robot that can be set to fire autonomously, but that operates in a very limited environment, is the robotic gun turrets deployed by South Korea on its side of the demilitarized zone (DMZ) between it and North Korea. The robot can be set to identify, track, and engage targets, including human targets, without human intervention. So although it is by some definitions a "lethal autonomous robot," it operates only within a unique and highly bounded environment, the Korean DMZ. Likewise, the Roomba®, a commercial vacuum cleaning device, operates effectively only within the floor of a home, and bomb disposal robots operate within a bounded space in much larger battlefields.

Such robots represent state-of-the-art expert systems, performing well within the space they are designed for, but with performance rapidly degrading outside those boundaries. They are thus unlike humans, who exhibit a general capacity to manage new conditions and environments. Expert systems perform poorly if at all beyond their design space; general intelligence systems are more able to adapt beyond their familiar boundaries. Also note that autonomy in a robotic system does not (yet) imply sentience, independent agency, or free will—themselves difficult and contentious concepts. These concepts are, however, important with regard to a serious component of the laws of war: the need to be able to assign responsibility for military actions to ensure compliance.

---

[1] In military usage, a "drone" is a vehicle used for target practice, as opposed to an "unmanned aerial vehicle" (UAV) used for surveillance and combat missions, but in common use all such devices are called "drones."

Just as "robot" has a multiplicity of definitions, the relationship between human and robotic systems is not nearly as clear as implied by terms such as "autonomous." Agreement on definitions remains elusive and subject to manipulation by activists pursuing utopian or dystopian agendas. There is instead a spectrum of human involvement ranging from human-controlled to human-supervised, all the way up to fully autonomous decision making and action.

Most currently deployed military systems are able to maneuver and perhaps even target autonomously, but require approval to fire. These include some torpedoes, many military drones, and Israel's Harpy missile system. Others have fully autonomous modes and can fire without human authorization, but only under limited circumstances. Samsung SGR-A1 robots deployed in the Korean DMZ, Navy Phalanx systems, and Patriot missiles are closer to this model. Fully autonomous systems in the near future are unlikely, and even here definitional questions arise: if a robot has wide discretion in target selection but is bounded by applicable rules of engagement, programmed ethics modules, and the laws of armed conflict, is it fully autonomous?

Why Military Robot Technology Will Advance

In 2001, Congress set as a goal for the U.S. military that one-third of American operational deep-strike aircraft would be unmanned within 10 years, and that one-third of all American ground combat vehicles were to be unmanned in 15 years. While by mid-2014 the U.S. Air Force was training more drone pilots than fighter and bomber pilots combined, these goals are still aspirational. The main reason is that current drones are not yet capable of operations in contested airspace, and development of UAV technology, while advancing rapidly, is not yet adequate for autonomous operation in uncontrolled environments. Nonetheless, robot technologies of all types are going to be widely deployed, at least by the United States, in conflict situations. This begs a deeper question: why is there interest in military and security robots at all?

There are five drivers behind the use of robotic systems in combat. The first is force multiplication: deploying robots can reduce the total number of soldiers needed. This factor becomes increasingly important as populations age, especially in developed countries, so that proportionately fewer young people are available to serve. It is no coincidence that the militaries most interested in robotics are those that are looking at upcoming demographic choke points, where low domestic birth rates and low levels of immigration signal potential difficulty in recruiting adequate military personnel in future. Beyond demographics, this factor favors increasing the degree of autonomy for each system, since continuous and labor-intensive human oversight diminishes the force multiplication actually achievable. For example, a large team is required to field a UAV today even though a traditional pilot is not required; efforts to reduce the personnel requirement per aircraft are underway. Autonomous systems, properly designed and deployed, have obvious potential to reduce personnel requirements without affecting force projection.

Second, robots expand the battlefield: they allow combat to be conducted over larger geographical areas. They extend the warfighter's reach, allowing soldiers to strike enemies from ever more distant and safer locales. This may not be the choice of the military deploying the robot but forced by the circumstances of the conflict. A diffuse global jihadist movement, for instance, may require a concomitantly global response. If cyber bots (the software robots mentioned at the beginning of this chapter) are considered to be robots, of course, geography becomes even less relevant: cyber is deployed based on network configuration and bot design, rather than any geographic requirement per se.

Third, robots reduce friendly casualties: they keep human fighters out of harm's way. This is critical for today's militaries, which face a sea change in public acceptance of warrior casualties. In World War I, for example, the First Battle of the Marne resulted in some 500,000 casualties on both sides; the First Battle of Ypres, approximately 200,000; and the Battle of Verdun, an estimated 700,000 to 900,000 casualties. In the Vietnam War, the United States suffered some 47,000 combat deaths in total. Combat deaths in Afghanistan totaled a little over 1,700, and in Iraq 3,500—significantly less than even in Vietnam, especially given how long the latter conflicts have lasted. The reality is that American public opinion will not accept significant American combat casualties; the same dynamic is evident in Europe[2] and can be observed to some extent even in Russia.[3] This is a significant constraint to a military charged with projecting national power, and creates enormous pressure to increase reliance on ever-more-autonomous robotic systems (as well as other alternatives, such as private military contractors).

Fourth, unenhanced human cognition is inadequate to respond effectively to the accelerating complexity and information density of the combat environment itself. This limitation has two implications. First, the cognitive function of the soldier must be distributed across technological networks rather than relying on the individual's sensory and cognitive systems. Robots will often be part of the resulting techno-human network, and provide the speed

---

[2] Relatively small numbers of casualties reinforced public pressure to withdraw European militaries from Iraq and Afghanistan.

[3] Soldier's families have expressed concern regarding Russian military casualties in the Crimea campaign.

and scope of information processing necessary in modern conflict environments. Second, in many environments robots are simply essential for effective action. Cyberattacks, for example, are so rapid and complex that no human could respond effectively before significant damage occurred. Both of these factors will lead to more cognitive processing power being designed into robotic systems, and drive development and deployment of increased (albeit still probably bounded) autonomy for technological systems. Note that this is not necessarily a choice the military would make; it is a choice made for the military by the conditions under which it must function. Robots cannot eliminate the Clausewitzean "fog of war," but because that fog is getting denser in modern conflict, they become necessary just to stay in the same place.

And finally, robots have the potential to save lives. "Dumb" weapons usually compensate for lack of finesse by being larger and more destructive. A "smart" bomb directed to a specific target can be smaller while still accomplishing its mission. This means that, properly designed and deployed, more sophisticated robotic systems have the potential to perform mission function with less collateral damage to non-combatants and civilian infrastructures.

In sum, the goal of military robotics is less to replace the human warfighter, and more to maintain, and where possible extend, the capability, scale, and effectiveness of the human fighting force, while reducing casualties among friendly forces and non-combatants. The powerful incentives for development and deployment of robotic systems faced by virtually all militaries should not be expected to wane, and it would be naïve to assume that robots will not be an increasingly important military technology.

Ethics and Autonomy in Robotic Systems

As with any technology, a modern military's use of robotic systems of all kinds in conflict situations must be done within applicable ethical and legal frameworks. This is not simply a matter of morality, but of self-interest. Failures to comply with ethical and legal standards in military operations often have serious consequences, both politically and pragmatically. Consider, for example, the effect of the My Lai massacre in Vietnam on American domestic political support for conflict, or the effect of the Abu Ghraib prisoner abuse controversy in Iraq, with the concomitant damage to America's public image worldwide. The damage to soft power such incidents create is substantial, and far outweighs the immediate tactical or operational gains, if any; this is especially true in a counter-insurgency environment.

Certainly, then, as militaries continue the development and deployment of intelligent autonomous weaponized robots at the current rapid pace, such systems should be deployed ethically, in a manner consistent with both standing and mission-specific rules of engagement (ROEs) and other legal and ethical constraints.[4] At least 47 nations are developing robots for the battlefield, which makes some form of "autonomous lethality" seem likely. This probability has led to the recognition that programming

---

[4] Note that this does not mean that all robotic systems should or must obey the laws of armed conflict. Modern battlefields are occupied simultaneously by espionage agencies, non-governmental organizations, private military firms, and other entities, all of which may be using similar robotic technologies, but may be governed by very different legal structures. ROEs are directives that tell units and individual warriors how they can use force and what actions they can take in their operations.

robots to make autonomous firing decisions raises some important ethical concerns.

These concerns have been recognized not just by critics, but by the militaries involved. For instance, the United States Air Force summarized the situation as follows:

> *Authorizing a machine to make lethal combat decisions is contingent upon political and military leaders resolving legal and ethical questions.... Ethical discussions and policy decisions must take place in the near term... rather than allowing the development to take its own path apart from this critical guidance.*[5]

In October 2010, Reuters[6] reported:

> *In a report to the U.N. General Assembly human rights committee, Christof Heyns said such systems [an apparent reference to U.S. drones that strike suspected Islamist militants] raised "serious concerns that have been almost entirely unexamined by human rights or humanitarian actors."*
>
> *"The international community urgently needs to address the legal, political, ethical and moral implications of the development of lethal robotic technologies," said Heyns, U.N. special rapporteur on extrajudicial executions.*

Human Rights Watch and other NGOs have mounted campaigns against so-called "killer robots," demanding precautionary bans on the technology. Those leery of lethal autonomous robots fear that they will simply refuse orders or malfunction, eventually escalating into a "robots run amok" scenario reminiscent of science fiction stories. Others fear that robots will always be incapable of making

---

[5] United States Air Force, *Unmanned Aircraft Systems Flight Plan 2009-2047* (Washington, DC: Headquarters, USAF, 2009), p. 41.

[6] Patrick Worsnip, "U.N. urged to set up panel on ethics of robot weapons," *Reuters* (21 Sept. 2012), available online: www.reuters.com/article/2010/10/22/us-un-rights-robots-idUSTRE69L5RL20101022.

complex decisions in battlespaces, and thus of following the requirements of the laws of armed conflict. They argue, for example, that robots simply cannot be programmed to discriminate appropriately between a combatant and a non-combatant.

Even if this challenge can be overcome, critics argue that robot warriors will give political leaders the illusion of being able to conduct "risk-free" warfare. This may lead governments to engage more readily in armed conflict, and to hide the nature and costs of that conflict from the public. (This issue is addressed in more depth in Chapter 2.) Other concerns include the probability of hacking, or misappropriation of the technology by terrorist groups. These fears are not unrealistic: researchers at the University of Texas at Austin have demonstrated the ability to take over a UAV with an unencrypted GPS system, alter its flight path, and even land it.

Most major military powers are developing and deploying robotic technologies of one sort or another. Some feel that the only possible prohibition of lethal autonomy may come from international treaties. Others suggest that prohibition is unnecessary because future autonomous robots may be able to perform better than humans under battlefield conditions. The eventual development and use of a broad range of robotic sensors will give machines better equipment for battlefield observations and performance than humans currently possess. Even now, one advantage of robotic systems is that they are able to integrate larger quantities of information, from more sources, far faster than a human possibly could in real time. Moreover, robots have the ability to act conservatively (i.e., they do not need to protect themselves in cases of low certainty of target identification).

An implicit assumption of the anti-robot activists is that humans generally comply with all relevant laws and norms, even under the stress of combat, an assumption which real world experience belies. Robots, on the other hand, could potentially be designed without the emotions that often cloud the judgment of human combatants, tempting them to commit war crimes. The argument is therefore that robots will have the benefit of not being susceptible to stress or psychological trauma, and, unlike human combatants, will not act out of anger, panic, or fear, and that they can in fact be designed and program to avoid ethical violations.

The Question of Responsibility

In addition to the arguments listed above, a few specific and more subtle issues arise around lethal autonomous robots. One of these is the responsibility for the wrongful deaths of non-combatants or other legal or ethical transgressions. Perceived infractions of international humanitarian law by unmanned systems could result in war crime charges, political fallout, a negative effect on troop morale, hostility among the local population, and citizen reticence toward mission accomplishment.

To some extent, however, this concern reflects a category mistake. War crimes are intentional acts that violate established law, and robots, certainly in the foreseeable future, will not be capable of intentionality. The appropriate analogy here is a soldier blaming the bullet he deliberately fired at a civilian for the war crime; no one is going to hold the bullet responsible. Similarly, a robot, just like any other military technology, may effectuate a war crime, but it is the individual who is responsible for the robot, not the robot, who is liable for the act.

The real question here is how to determine responsibility with a complicated and, at the beginning, novel piece of machinery, rather than trying to allocate responsibility in whole or in part to the machinery. In an environment of rapid technological evolution, this raises a question of how the relationships among design and production systems, military personnel and institutions, and their robotic tools, can be managed. Such questions are not insurmountable: every responsible military has long experience with assigning responsibility for operations—and failures—across the chain of command.

Regarding the very real challenge of combatant identification and distinction, a number of measures can be used to limit or prevent collateral damage, and to establish responsibility. First and foremost, robots could be employed, at least initially, in limited circumstances where the likelihood of civilian encounters can be minimized and where the situation is bounded enough that existing artificial intelligence (AI) capabilities can be used. Examples of these "bounded" or highly scripted situations include room clearing, counter sniper operations, and DMZ or perimeter protection, with individual systems designed for specific situations. Robots can be programmed to perform a situational analysis whereby a number of criteria must be met before a shot can be fired. If the robot finds itself in an unexpected situation, it can be programmed to merely observe it.

Humans in the Loop

While keeping humans involved in the decisions made by robots—particularly kill decisions—may appear to alleviate these problems, the reality raises many more questions. The most fundamental question, of course, is whether human judgment is better than properly framed and governed machine judgment. Many critics of military robots assume that it is, but military experience (which many critics lack) questions that assumption. For example, the U.S. Navy guided missile cruiser USS *Vincennes* shot down a civilian Iranian airliner over the Persian Gulf on July 3, 1988, with a loss of 290 lives. The *Vincennes* was equipped with the Aegis robotic fire control system; the system correctly identified the flight as civilian, but the crew on the bridge, apparently as a result of a psychological condition called "scenario fulfillment," a form of groupthink, fired surface-to-air missiles in the mistaken belief that they were under attack from a hostile warplane.

In many military systems, the human is the component with the "lowest bandwidth"—i.e., slowest rate of information processing. Hence, there is a tension between limited human cognitive ability and ensuring that optimal battlefield decisions are made, especially as the battlefield environment becomes more complex and information-dense. That is, "sufficient" information for a human to make an informed decision is possibly "too much" information for the human to process in a reasonable amount of time. This condition may be especially true in modern warfare, when intense and rapid attack may require the information processing power of robotic systems to keep up with the challenges. A human is simply incapable of the required speed and information processing capacity. Thus even with humans in the loop, it may be difficult to establish real accountability.

Another corollary concern about keeping humans in the loop arises from remotely piloted UAVs. Some have likened these military operations to video games, charging that remote pilots have a "Playstation mentality" and that killing at a distance encourages escalated violence. These assertions are mostly made by people who have no direct experience, and are thus far unsubstantiated. Indeed, all data indicate the opposite: operators have been known to suffer from exhaustion and post-traumatic stress disorder. Other questions are more germane. For instance, are uniformed UAV operators in Nevada valid enemy targets? More broadly, as weapons and technology systems become less geographically bounded, do they in and of themselves extend the battlefield deeper into civilian systems, and to new regions? The laws of war seem to imply that they do.

Despite that fact that the Air Force is now graduating more UAV pilots than manned pilots, operating unmanned systems is more labor-intensive (at least for now). It is thus not unlikely, given trends towards privatizing military operations of all kinds, that at least some armed forces will close labor gaps by employing civilian or private military operators, especially as advanced technologies are adopted by militaries that do not have the skill sets required to operate such systems.

However, having non-uniformed personnel involved in operations that frequently end with lethal attacks on human targets is problematic because different legal regimes may apply. Non-military operators could conceivably be found guilty of murder, at least under current law. In a worst-case scenario, if there is no statute of limitations on murder and a former UAV operator is traveling in a nation in which military operations have been conducted, it is possible that the individual could be charged and tried. This is not a question unique to robotic systems, since it results from the privatization of military operations in general.[7]

Context and Technology

Many of the issues arising from robotic technology involve the questions of boundary and context. A technology, such as autonomous shoot-to-kill capability, that would be unethical if deployed in the confusion of an urban battleground, may be competent and lawful in the context of a DMZ. Questions of responsibility are limited

---

[7] The same issue arises for private military contractors loading bombs into weapon systems closer to the battlefield, for example.

if applicable rules of engagement control deployment to those situations where robotic responses are predictable and bounded. Such bounds can arise physically (e.g., static robots deployed in a DMZ environment), operationally (e.g., de-mining robots are deployed only when de-mining activities are required), or according to mission (robots may be deployed in purely military environments, but not in policing environments). Some robotic technologies may not be deployed because bounds are not assured. For example, software bots that might have crippled Iraq's financial system were allegedly not deployed prior to the U.S. attack in March 2003 because it was unclear that their activities could be limited to the target, given the global interconnection of financial networks.

In the near term at least, most robots will likely be deployed alongside, and not as a replacement for, human combatants. Under some circumstances, this might lead to new dynamics and behaviors. Robots participating on an integrated team combining human soldiers and autonomous systems could have the capability of monitoring ethical behavior in the battlefield by all parties and reporting infractions.[8] Some have suggested developing these robots, perhaps even requesting advice from military legal experts when needed ("lawyer in the loop"). Such a device could, however, change the robot from a perceived team member to a perceived enemy, with unpredictable results. And if such robots were hackable, the potential for sophisticated "lawfare"—using the laws of war against the military organization trying to comply with them—is obvious.

Some experts offer another scenario. Robotic platforms offer at least the potential to reduce collateral damage by identifying legitimate targets and the least damaging way

[8] Tasers, for example, already come equipped with a "black box" that records the date, time, and duration of each use.

to attack them. Under such circumstances, using robotic systems might not be merely acceptable; it might become a moral imperative.

Psychological dimensions must also be considered. Experience in Iraq indicates that soldiers in mixed human/robot teams tend to become attached to their companion robots.[9] When they are damaged, for example, soldiers do not want new ones, they want the old ones to be repaired. In some cases, soldiers have even come to embrace a doctrine of "no robot left behind." For soldiers, developing strong bonds with robots that are supposed to protect them (e.g., minesweepers) could ultimately get them killed. So designers are faced with the problem of making the robots likeable, but not *too* likeable.

Conclusion

Many research questions regarding lethality and autonomy of robotic systems have yet to be resolved. Operational issues associated with the rapid technological evolution and deployment of such systems are complex, and the challenge posed by robotic technologies to the existing norms and laws of armed conflict are both real and difficult. Discussion is complicated by the rapid and unpredictable acceleration of relevant technologies and the overlap between civilian applications, private firms, related fields such as artificial intelligence, and the military and security domains. How, for example, will the rapid evolution of "autonomous" vehicles in the civilian world, or the integration of AI and data-mining technologies, affect the design of military robotic systems?

Although robotics, defined broadly, is a complex and very emotive space, roboticists, political leaders, and citi-

---

[9] This is a familiar psychological response to robots in civilian applications as well.

zens alike should not ignore these difficult ethical issues. This is particularly vital as the technologies spread into armed forces with different worldviews and norms, into private or quasi-private military firms and institutions, and broadly into civil society.

ABOUT THE AUTHOR

Braden R. Allenby

Braden R. Allenby is President's Professor of Civil, Environmental, and Sustainable Engineering, and the Lincoln Professor of Engineering and Ethics, at the School of Sustainable Engineering and the Built Environment at Arizona State University, in Tempe, Arizona. He is also the Founding Chair of the Consortium for Emerging Technologies, Military Operations, and National Security, as well as the Founding Director of the Center for Earth Systems Engineering and Management, at ASU.

Prior to joining ASU in 2004, he spent 20 years in senior positions with AT&T, responsible for the company's global environmental, health, and safety activities.

Dr. Allenby received his BA from Yale University, his JD and MA (economics) from the University of Virginia, and his MS and Ph.D. in Environmental Sciences from Rutgers University. He is an AAAS Fellow and a Fellow of the Royal Society for the Arts, Manufactures & Commerce.

Dr. Allenby's most recent book is the edited volume *The Applied Ethics of Emerging Military and Security Technologies* (Ashgate Press, 2015). His other books include *Industrial Ecology and Sustainable Engineering* (co-authored with Tom Graedel in 2009), *The Techno-Human Condition* (co-authored with Daniel Sarewitz in 2011), *The Theory and Practice of Sustainable Engineering* (2012), and *Innovative Governance Models for Emerging Technologies* (edited volume with Gary Marchant and Ken Abbott, 2013).

ACKNOWLEDGEMENTS

The author would like to thank the Lincoln Center for Applied Ethics at Arizona State University for their support for his work in the applied ethics of emerging military and security technologies, and the New America Foundation/Arizona State University Center for the Future of War, and the Arizona State University School for Sustainable Engineering and the Built Environment, for supporting his research in this domain. Thanks also to G. Pascal Zachary for conceiving this project, and to Jason Lloyd for his editing efforts.

The five essays in this volume represent perspectives developed by the author based on his experience working as a government and corporate attorney; an executive with global responsibilities for environment, safety, and sustainable design for a Fortune 500 company; a student of emerging military and security technologies; a Stockdale Fellow at the U.S. Naval Academy; and the founding chair of the Consortium for Emerging Technology, Military Operations, and National Security.

All material reflects the author's perspectives, and does not represent the viewpoint of any of the institutions with which the author has been or are now associated. In many cases, the chapters draw from previous works as noted below:

- The introduction was written specifically for this volume.

- Chapter 1 is loosely based on an article prepared for the law review journal *Jurimetrics* (in press). Used by permission.
- Chapter 2 is loosely based on the introductory chapter by B. R. Allenby, *The Applied Ethics of Military and Security Technologies,* edited by B. R. Allenby (Farnam, UK: Ashgate Press, 2015), pp. xiii-xxxii. Used by permission.
- Chapter 3 is based on B. R. Allenby and M. Hagerott, "Universal Conscription as Technology Policy," *Issues in Science and Technology* 30, no. 2 (2014): pp. 41-46. Used by permission.
- Chapter 4 was written for this volume, and is loosely based on excerpts from C. S. Mattick, B. R. Allenby, and G. R. Lucas, Jr., *Chautauqua Council Final Report: Implications of Emerging Military/Security Technologies for the Laws of War* (Tempe, AZ: Arizona State University Lincoln Center for Applied Ethics, 2012).
- Chapter 5 was written for this volume, and is loosely based on excerpts from C. S. Mattick, B. R. Allenby, and G. R. Lucas, Jr., *Chautauqua Council Final Report: Implications of Emerging Military/Security Technologies for the Laws of War* (Tempe, AZ: Arizona State University Lincoln Center for Applied Ethics, 2012).

Made in the USA
Lexington, KY
11 November 2017